D1742236

acer/
7

BINNIE & PARTNERS
CONSULTING ≈ ENGINEERS

PROCEEDINGS OF THE INSTITUTION OF CIVIL ENGINEERS

Greater Cairo Wastewater Project

SUPPLEMENT TO CIVIL ENGINEERING

The publishers are grateful to Acer Consultants Limited and Binnie & Partners whose sponsorship has made this special issue so comprehensive

CONTENTS

Proc. Instn
Civ. Engrs,
Civ. Engng
Greater Cairo
Wastewater
Project,
1993, 2

FOREWORDS

Baroness Chalker of Wallasey

I am delighted to contribute a foreword to this special issue, which reviews the purpose and progress of a major project in public health engineering.

Most of us have become so conditioned to life in a complex urban environment that we take a great deal for granted. Only when an event such as drought or a severe storm causes temporary inconvenience are we reminded of the highly sophisticated and ingenious systems which constantly maintain our cosy existence. So reliable and well-built are our services of water supply, power, waste disposal and drainage that we seldom think of the effort and investment that goes into their provision and operation.

In the developing world the situation is very different. Populations continue to grow at a frightening pace, and the pattern of economic development encourages an increasing number of people to move from their former rural existence into the cities. Those cites therefore expand very rapidly, placing an intolerable burden upon the basic infrastructure, which then collapses under the load. Such situations clearly threaten the health and well-being of the inhabitants, but the cure is expensive.

In the case of Cairo, with its roughly 12 million inhabitants, the wastewater system had to be upgraded from a design capacity adequate to serve only about one million people. This huge task has been tackled by means of co-operation between a number of donors including the United States, the United Kingdom, the European Investment Bank, the European Community, and by commercial loans. British engineers and companies have played a prominent role in the work of both design and construction. We are proud of the contribution which they have made to this vital undertaking, and to many others as well.

Courtesy ODA

The Rt Hon.
Baroness Chalker of
Wallesey is Minister
of State for Foreign
and Commonwealth
Affairs and Minister
for Overseas
Development.

H.E. Eng. Hassaballah El-Kafrawi

The Greater Cairo Wastewater Project is a remarkable engineering achievement which has produced significant improvements to the environment and public health of the inhabitants of the city.

The project is also significant as an example of international co-operation in the successful implementation of major infrastructure works. The original arrangements involved the collaboration of Egypt, the USA and the UK in the financing and engineering of the project. More recently, there has been involvement from Italy and the European Community.

It is fitting that the significant part played by British engineers and companies, working in association with their Egyptian colleagues, should be recorded in this publication. I wish to express the appreciation of the Egyptian Government for their contribution to the well-being of the citizens of Cairo.

His Excellency
Eng. Hassaballah
El-Kafrawi
is Minister of
Reconstruction,
New Communities,
Housing and
Public Utilities

Introduction and historical overview

D. G. M. Roberts, CBE, FEng, MA, FICE, FIMechE, Eng. S. A. Salem, BSc, and E. W. Flaxman, FEng, BSc(Eng), DIC, FICE

*Proc. Instn
Civ. Engrs,
Civ. Engng,
Greater Cairo
Wastewater
Project,
1993, 3–7*

Paper 10227

■ Introduction

A wastewater scheme for the largest, and fastest growing, city in Africa is bound to be a major undertaking. In 1985, a series of seven papers describing the first eight years of the Cairo Wastewater Project was presented to the Institution of Civil Engineers (ICE),[1-7] and other technical papers have been published since then in Egypt and in the international press.[8]

2. In January 1992 a very large element of the new sewerage system on the East Bank of the River Nile was formally opened by President Mubarak, accompanied by His Royal Highness the Duke of Gloucester. Towards the end of 1992 a substantial part of the new sewerage system on the West Bank was successfully commissioned. The present series of papers is being published to bring the account up to date.

3. This Paper sets out the historical perspective of the scheme. Fig. 1 shows the extent of the urban area of Greater Cairo at the time the project was about to start. Earlier wastewater schemes are then described briefly, followed by an account of the serious problems posed by overloading of the facilities in the 1970s and early 1980s. The Paper concludes with a short account of the way in which the present project was initiated by the Egyptian government and a master plan developed.

4. For many centuries after its establishment as the capital of Egypt, the city grew almost exclusively on the East Bank of the River Nile. Until the first major road bridge was constructed in 1872, the business centre, industry and the great majority of the population were all located on the East Bank.

Genesis of the first sewerage scheme

5. Sanitary conditions in Cairo began to give rise to concern in the last quarter of the nineteenth century. In 1885 a commission was established to study the best means of improving the worsening situation.

6. In 1889 Mr Baldwin Latham was engaged by the Egyptian government to prepare a scheme for the drainage of the city. He reported in 1890 and put forward proposals catering for a population of 500 000 in an area of 2500 ha by means of a pumped system using 28 pumping stations. An international competition was then held, which resulted in the submission of 30 designs. An international commission was then appointed to consider these submissions; it was unable to recommend any of them.

7. Further proposals were prepared in 1893 by Mr J. Barois, who criticized the use of subsidiary pumping stations and instead advocated a wholly gravitational system draining to a single major pumping station. Sir William Willcocks also prepared a scheme in 1899, but only minimal construction followed either of these reports.

8. When Mr Carkeet James was appointed to prepare a drainage scheme in 1906 he found that sanitary conditions were 'very defective'. There were virtually no drains in the city and in an important paper to the ICE in 1916 he recorded having had the opportunity 'which does not often now present itself to an engineer, to design a modern scheme of drainage *de novo*'.[9]

9. Fortunately for his successors, Carkeet James adopted a general layout for the sewerage system which accorded well with the topography of the city and has stood the test of time. Between the limestone hills to the east, on which the citadel was founded, and the plateau to the west where the Pyramids were built, most of the city is on the flood plain of the River Nile and is very flat, with a slight general fall towards the north. Carkeet James reverted to the widespread use of small ejector pumping stations, and directed the 1·6 m dia. main collector sewer towards the north-east.

10. The collector terminated at Ein Shams, where a major steam-driven pumping station was constructed to convey the sewage a further 12 km to the north. There, limited purification facilities and a sewage farm of 1260 ha were established at Gabal el Asfar. The scheme was designed for a future population of just under one million. Because significant rainfall is relatively rare in Cairo—the annual precipitation averages about 25 mm—only a small allowance had to be made for surface water entering the sewers.

11. By the time Carkeet James presented his paper to the ICE in 1916, 64 km of small diameter clayware sewer had been laid, 63 ejector stations constructed in the central area of the city, the 14 km main collector sewer was complete, the pumping station at Ein Shams was in operation, and sewage from the properties already connected to the system was being conveyed to the farm where 40 000 trees had been planted.

Lessons learnt from the first scheme

12. Construction of the scheme confirmed the general difficulty of laying deep sewers in the soft, waterlogged alluvium of the Nile Valley. Carkeet James limited the depth of excavation for small sewers to 5 m in wide streets and 3 m in narrow streets. Some of the deeper ejector stations had to be constructed under compressed air, and excavation for the main collector encountered much greater quantities

*D. G. M. Roberts,
Life President,
Acer Group
Limited; Board of
Control, AMBRIC
(1979–92)*

*Eng. Salama A.
Salem, Chairman
and Member of
Board of
Management of
CWO*

*E. W. Flaxman,
Consultant,
Binnie & Partners;
Board of Control,
AMBRIC
(1979–89)*

of groundwater than had been expected. The 14 km length of collector took four years to construct, from 1910 to 1913, and at one time required the use of 27 steam-driven pumps. One tonne of coal was consumed for every 10 m³ of excavation and over 3795 m³ of timber had to be left in the trenches. Dewatering of the excavation for the main pumping station at Ein Shams was carried out using ten pumps and thirteen boilers; the total quantity of coal consumed was 7100 t.

13. The cost of the scheme was almost 2 million Egyptian pounds (£E), paid for initially by the Egyptian Government. In 1912, house-tax in the city was raised from 8·5% to 10% to help defray the cost. The amount raised annually from this source was expected to be £E40 000 or more, but as the annual cost of maintenance amounted to £E32 000, hopes for financial viability had to be pinned on profits

from the sewage farm, which were estimated at £E30 000 per annum.

14. Most of the scheme functioned well. The smaller sewers and many of the sewage ejector stations were still operating 70 years later with minimal need for attention. However, a major problem arose owing to corrosion of the main collector. This collector was constructed of mass concrete and serious corrosion above the water line was discovered in 1918, which was due to the generation of hydrogen sulphide in the warm conditions, the long retention of sewage in the system and the difficulties of providing ventilation. Various remedies were proposed, but none proved feasible because the collector had to be kept in service to cater for the steadily increasing flow of sewage.

15. When increasing population and flows necessitated construction of a second, larger, main collector in 1925 it was lined with high-

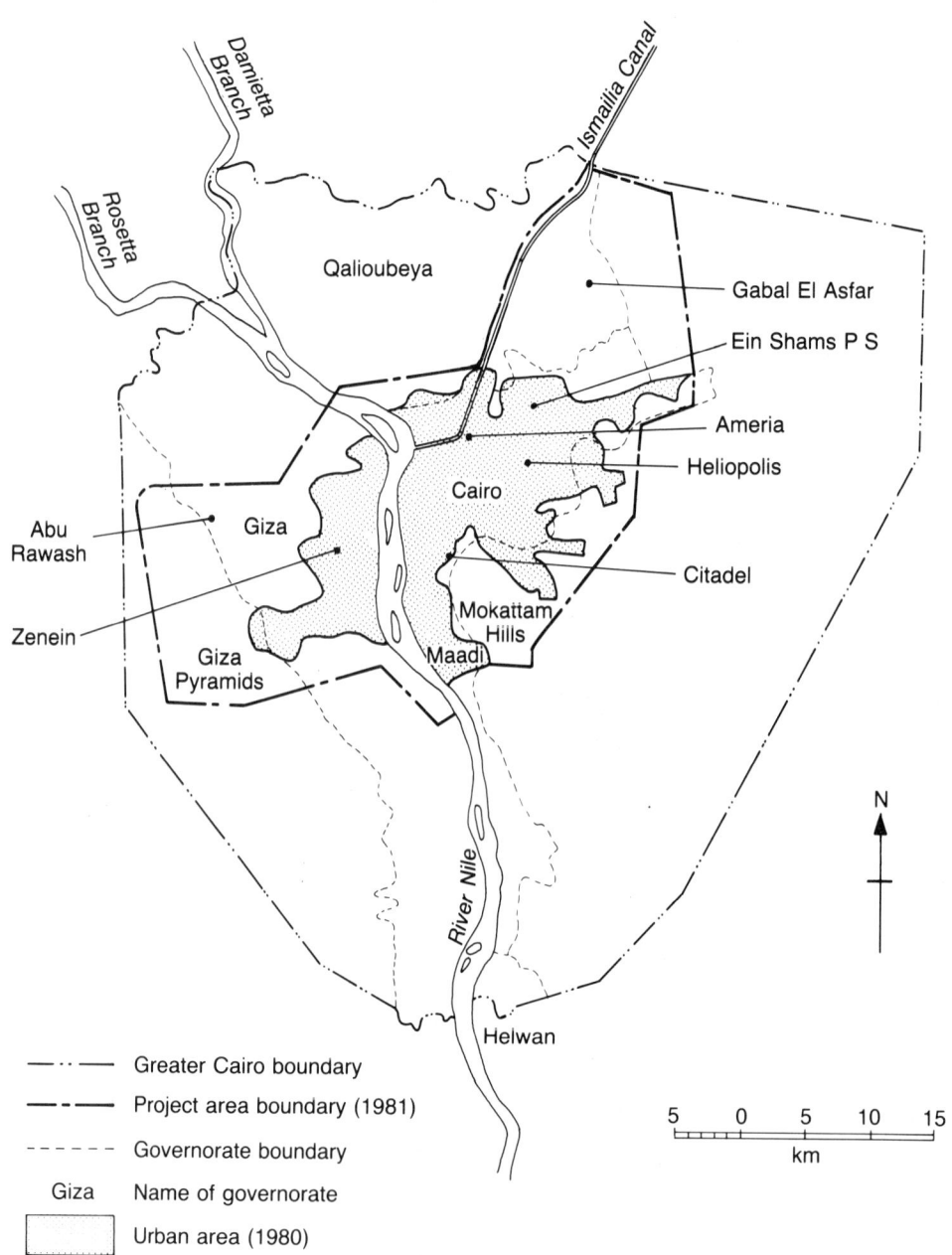

Fig. 1. Greater Cairo area

Damietta Branch

Rosetta Branch

Ismailia Canal

Qalioubeya

Gabal El Asfar

Ein Shams P S

Ameria

Heliopolis

Cairo

Abu Rawash

Giza

Citadel

Zenein

Mokattam Hills

Giza Pyramids

Maadi

River Nile

Helwan

N

— ·· — Greater Cairo boundary

— · — Project area boundary (1981)

- - - - - Governorate boundary

Giza Name of governorate

☐ Urban area (1980)

5 0 5 10 15
km

quality blue bricks set in cement mortar and with joints pointed with tar and pitch. It proved thoroughly serviceable, and when a third collector had to be added in the 1950s it, too, was brick-lined.

16. Deposition of sand in the sewerage system also proved to be a problem. When Carkeet James's paper had been discussed at the ICE much attention had been directed to the design velocities and gradients adopted for the sewers. Most of those who took part in the discussion, including several authorities of the day from the UK and the USA, were of the opinion that lesser gradients and velocities could have been adopted, with a significant saving in depth, and thus of excavation costs.

17. In his reply, Carkeet James made the important point that sewage on the East Bank of the River Nile was 'heavily laden with solid matter' due to the cleansing of cooking utensils with earth or sand. Although it was not until many years later that engineers began to develop a theoretical basis for calculating appropriate velocities in sewers, Carkeet James clearly appreciated the vital importance of the sand content of sewage on the velocities likely to be needed if deposition was to be avoided.

18. The provision of the second main collector, a new pumping station at Ameria and a further pumping main to Gabal el Asfar were described in a paper presented to the ICE by Pinson in 1930.[10] Slightly higher velocities of flow were adopted for this second collector. Despite this, grit deposition continued to occur. When, in 1935, Mr Godfrey Taylor was invited to visit Cairo and report on proposals for improvement, one of the main problems occurring was grit settlement in the main collectors.[11]

Growth of the city and its effects

19. When Carkeet James designed the first sewerage scheme in 1907 the population was 658 000. That already showed a significant increase over Mr Latham's figure of 500 000 in 1889 and he allowed for the population to increase over the next 25 years to 960 000 in 1932.

20. A census carried out in 1927 showed the population to be 1 064 000. Since then, the modern history of Cairo has been dominated by population growth. The census figures tell their own tale

1937	2 020 000
1947	2 960 000
1960	4 830 000
1966	6 150 000
1976	7 970 000
1986	10 660 000

21. At the time of writing, the population of Cairo is estimated to be about 13 million. The number is still increasing at a greater rate than was experienced in either London or New York during the decades of peak growth during the nineteenth century, as shown in Fig. 2.

22. Added to the increase in population has been a massive increase in per capita water usage associated with much improved living standards. The design of the first sewerage scheme was based on an average per capita discharge of 50 l/day; the actual anticipated rates varied from 150 l/day for the more affluent to only 10 l/day from the majority of the inhabitants.

23. Since that time, large water supply schemes have been installed, drawing water from the River Nile. Average per capita usage is now estimated to be about 310 l/day.

24. These increases in both population and water usage within the city soon gave rise to serious overloading of the wastewater system. The difficulties were compounded by wars. The Second World War delayed by 15 years construction of the third collector, and severe shortages of money for investment in infrastructure followed the subsequent conflict with Israel.

25. During the 1950s and 1960s various improvement schemes were carried out as money allowed. These included a crash programme of relief works known as the Hundred Days Scheme, the construction of a syphon under the Nile to convey flows from the East to the West Bank, and the provision of the first major activated sludge treatment plant at Zenein on the West Bank of the River Nile. However, the growth of the city and of wastewater flows outstripped all these measures.

Situation in the 1970s

26. By the middle of the 1970s the growing overload on the wastewater facilities had reached the state where sewage flooding occurred daily in streets at over 120 different locations. The risks to public health, to business, commerce and tourism were obvious.

27. Staff carrying out maintenance and repair were increasingly handicapped by a shortage of spare parts. The vastly increased sewage flows had swamped many of the treatment facilities and a large volume generated on

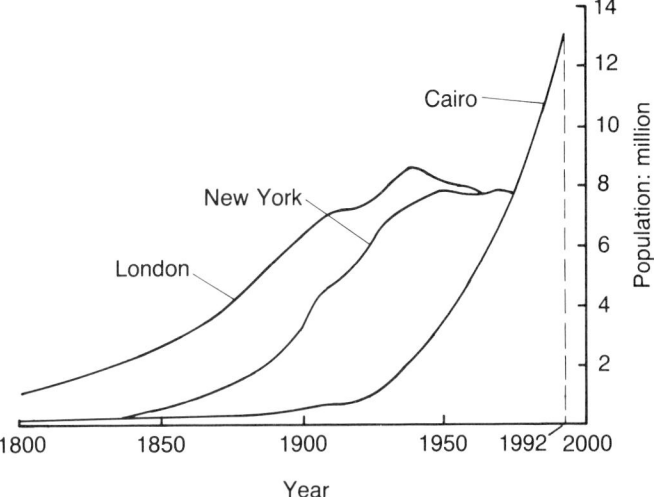

Fig. 2. Population growth of cities

the East Bank was having to be discharged into an irrigation drain through an emergency canal constructed many years earlier. The activated sludge plant installed on the West Bank was progressively made ineffective by the irreparable breakdown of grit removal facilities, followed by failure of the sludge return pumps.

28. The threat posed by the steadily deteriorating wastewater situation made it imperative that something should be done. In 1974 the Ministry of Housing and Reconstruction (MOHR) formed a high-level advisory committee on reconstruction, composed of former ministers and very senior officials. One of the committee's first priorities was to tackle the wastewater problem.

Initiation of the present project

29. Terms of reference for a Greater Cairo Wastewater Study were drawn up and issued to 40 firms of international consulting engineers in 1975. The terms of reference called for a study of the existing situation, development of a master plan for long-term development of wastewater facilities, and the identification of top-priority projects which would alleviate the current problems. Concurrently, similar proposals for master plan studies were invited for Alexandria and Helwan, the mainly industrial town at the southern boundary of Cairo.

30. Early in 1976 proposals for the Cairo study were submitted to MOHR by 30 firms or groups of firms from seven different countries. Because the project would clearly be of major size and importance, several groups or consortia were formed by those submitting proposals. The firms of two of the authors, John Taylor & Sons (now Acer John Taylor) and Binnie & Partners (Overseas) Ltd, formed a new firm— Taylor Binnie and Partners (TBP)—and submitted their proposal in collaboration with Dr Abdel Warith of Cairo. Eventually, in November 1976, TBP concluded an agreement with MOHR for a study costing about £1 million, financed partly by the government of Egypt and partly by the Arab Fund based in Kuwait.

Master plan

31. The master plan report was submitted by TBP early in 1978.[1][2] The report confirmed the inadequacy and worsening condition of the existing wastewater facilities, owing principally to greatly increased population and lack of funds. Because of the very high density of population—averaging 275 persons per hectare over the whole city and over 1000 in some areas—the only practicable solution to the problem was extensive augmentation and extension of the water-borne sewerage system.

32. The master plan proposals followed the general layout of the existing system in continuing to drain most of the East Bank towards the north-east and most of the West Bank towards the west. Because of problems with traffic disruption, it was proposed that a new main collector and several other important new sewers on the East Bank should be constructed in tunnel.

33. Two large new sewage treatment works were proposed. That on the East Bank was to be on part of the site of the original sewage farm at Gabal el Asfar; that on the West Bank was to be at Abu Rawash, where a small facility had been established in the 1930s.

34. Consultation with planners and other government officials led to the adoption of a design horizon of 2000. At this date it was anticipated that the population of the study area would have increased to 13·9 million. This figure is somewhat less than that of Greater Cairo as a whole, since Helwan is excluded, but it is greater than the population of Cairo Governorate, as major parts of Giza and Qalioubeya Governorates are also covered.

35. Because of the seriousness of the situation, the top-priority projects identified were necessarily large and relatively costly. The main collector on the East Bank, for example, would be up to 5 m in diameter and would be constructed up to 26 m below ground level. Bearing in mind the difficulties encountered with groundwater at much shallower depths during the construction of the original scheme, such proposals were possible only because of much improved construction methods developed in recent years.

36. The recommendations of the TBP master plan report were accepted by the Government of Egypt in 1978. The aims of the project may be summarized as follows

(a) to remove wastewater from the city in a sanitary manner, making maximum use of existing facilities

(b) to treat the wastewater to a standard permitting reuse in agriculture.

37. The papers which follow in this special issue describe how the first aim has already been achieved throughout much of the city, and also the progress that is being made towards achieving the second aim.

References
1. ROBERTS D. G. M. and FLAXMAN E. W. Greater Cairo wastewater project: history, development and management philosophy. *Proc. Instn Civ. Engrs*, Part 1, 1985, **78**, Aug., 711–728.
2. KELL A. D. K. *et al.* Greater Cairo wastewater project: organisation, financial arrangements and contact strategy. *Proc. Instn Civ. Engrs*, Part 1, 1985, **78**, Aug., 729–743.
3. DARLING R. S. and DRAKE J. Greater Cairo wastewater project: studies and master plan. *Proc. Instn Civ. Engrs*, Part 1, 1985, **78**, Aug., 745–763.
4. HALL T. D. and GRAY M. W. Greater Cairo wastewater project: rehabilitation works. *Proc. Instn Civ. Engrs*, Part 1, 1985, **78**, Aug., 765–784.
5. SWAYNE T. A. and GRIMES J. F. Greater Cairo wastewater project: design of east bank stage 1 works pumping stations, culverts and treatment plant. *Proc. Instn Civ. Engrs*, Part 1, 1985, **78**, Aug., 785–805.

6. ELLIOTT I. H. and SPOLTON R. L. Greater Cairo wastewater project: design of east bank stage 1 works tunnels. *Proc. Instn Civ. Engrs*, Part 1, 1985, **78**, Aug., 807–829.

7. SKINNER W. D. and PRONSKE K. P. Greater Cairo wastewater project: design of west bank stage 1 works sewers, culverts and pumping stations. *Proc. Instn Civ. Engrs*, Part 1, 1985, **78**, Aug., 831–840.

8. Four papers presented at: *59th Annual Conference WPCF*. Los Angeles, USA, October 1986.

9. JAMES C. Main drainage of Cairo. *Min. Proc. Instn Civ. Engrs*, 1915–16, **202**, 57–139.

10. PINSON A. O. W. D. Cairo main drainage extensions. *Min. Proc. Instn Civ. Engrs*, 1930–31, **231**, 112–160.

11. JOHN TAYLOR AND SONS. *Report on Cairo main drainage*. JTS, 1936.

12. TAYLOR BINNIE & PARTNERS. *Wastewater master plan report*. TBP, London, 1978.

Proc. Instn
Civ. Engrs,
Civ. Engng,
Greater Cairo
Wastewater
Project,
1993, 8–17

Paper 10228

Project objectives, organization and implementation

A. D. K. Kell, BSc(Eng), MSc, DIC, MICE, MIWEM, *Eng. T. A. Saada,* BSc, *and J. P. Somerville,* BSc, MSc, FICE, FIWEM

A. D. K. Kell,
Partner, Binnie &
Partners; Board
of Control and
former Project
Director,
AMBRIC

Eng. M. Talaat,
Abou Saada, Vice
Chairman and
Member of Board
of Management of
CWO

J. P. Somerville,
Projects Group
Manager, Acer
Consultants
Limited; Board of
Control and
former Project
Director,
AMBRIC

■ Introduction

Following the completion of the Taylor Binnie and Partners (TBP) master plan in 1978 the Government of Egypt (GOE) sought to proceed with the implementation of the project. It had been expected that there would be continuing financial support from the Arab Fund but in the event this was not forthcoming. The GOE looked for other sources of foreign aid and obtained commitments of assistance from the UK Overseas Development Administration (ODA) and the US Agency for International Development (USAID). In line with the sources of funding it was agreed that the GOE would procure engineering services from firms of British and American consultants who would jointly act as Programme Manager and Engineer to the implementing authority, with support from local Egyptian consultants.

Project objectives

2. The terms of reference prepared for the consultants established primary goals for the project. These were to increase the capacity of Cairo's existing sewerage system by cleaning and rehabilitating as much as was economically practicable and to prepare a plan for the expansion of the system to meet the current and future wastewater collection and disposal needs within the Greater Cairo Project Area through to the year 2000.

3. The proposals were to be reviewed for environmental suitability, in particular having regard to

(a) the improvement of sanitary conditions within the project area and in areas affected by wastewaters generated in the project area

(b) the improvement and control of the quality of wastewater receiving waters so as to protect their beneficial uses, including agricultural irrigation, fishing and urban water supply and also their flora and fauna

(c) other environmental and socio-economic impacts of the project.

4. It was envisaged that the plan would need to have the following features

(a) an immediate programme to rehabilitate the existing collection system

(b) facilities which would relieve the overloading of the existing system and minimize the risk of wastewater flooding

(c) provision of a new collection and conveyance system to meet the growing needs of the city until at least the year 2000

(d) provision of treatment facilities to contemporary standards adopted in developed countries

(e) facilities which were of a good quality for long life, represented good value for money and made maximum use of local products and materials

(f) facilities which were easy and safe to operate and which could be sustained without high cost or effort

(g) facilities which would permit the beneficial reuse of treated wastewater and sludge at a later stage.

5. During 1979 arrangements were put in hand by the GOE and by the supporting foreign funding agencies for the implementation of a project to meet these objectives.

Government of Egypt's arrangements for implementation

6. To implement a project of the size and complexity envisaged it was necessary for the GOE to involve a large number of public sector organizations and to make some special arrangements.

7. The sponsoring ministry for the project is the Ministry of Reconstruction, New Communities, Housing and Utilities (MOHU). Since its inception the project has been under the personal guidance of the Minister, HE Eng. Hassaballa el Kafrawi.

8. Initially the organization responsible for the implementation was the General Organization for Sewerage and Sanitary Drainage (GOSSD) which was the authority responsible for Cairo's sewerage facilities. In 1981 the MOHU decided that the functions of GOSSD should be split, and created the Cairo Wastewater Organization (CWO), giving it authority for the execution of the new wastewater projects for Greater Cairo and for Helwan. The other organization, now known as the General Organization for Sanitary Drainage (GOSD), was given the responsibility for the operation and maintenance of the existing system and for capital works not forming part of the Greater Cairo and Helwan projects.

9. CWO does not have independent control of its budgets nor the authority to approve and award contracts. This is retained in the MOHU. Recommendations for contract award are referred to the Ministry's Higher Decision Committee for action. Conditions of contract are also subject to higher approval—these being referred to the Council of State for review before tenders are invited.

10. CWO is responsible for preparing its own budgets both for local and foreign currency and for submitting these to MOHU for consolidation into the Ministry's overall budget. Allocations of funds resulting from

budget requests are made by the Ministry of Planning for local and foreign currency. However, financing agreements for foreign currency through grants or loans are administered by the Ministry of International Co-operation.

11.　The project area of 875 km² lies within the boundaries of three governorates—those of Cairo, Giza and Qualioubeya (see fig. 1 in reference 1). It has been necessary for CWO to maintain close liaison with these organizations on many matters, including land acquisition and temporary occupation arrangements during construction. CWO has also needed to maintain close liaison with the Ministry of Public Works and Water Resources (MPWWR), particularly with respect to the discharges of sewage effluent into the irrigation drainage system.

12.　In 1986 MOHU decided to give further high level support to CWO by creating a Board of Directors comprising senior government officials having a wide range of backgrounds. Many of the members of the board are former government ministers. The board also has representatives of the Council of State, and of the major Egyptian water and wastewater authorities. H.E. the Minister has delegated his authority to this board. The arrangements for implementation made by the GOE are shown diagrammatically in Fig. 1.

Arrangements for funding and insurance

13.　During the course of 1978 discussions were held between the GOE, ODA and USAID concerning the initial funding of the project. The two foreign aid agencies agreed to contribute roughly similar amounts (£50 million and US$100 million) to fund the foreign currency requirements for the initial phase of construction and the associated consultancy services. The UK funds were to be earmarked for construction on the East Bank of the River Nile and the US funds for the West Bank and rehabilitation.

14.　One problem identified at this time was the Egyptian decennial liability requirements which could in certain circumstances make contractors and consultants liable for faults discovered up to ten years after the completion of construction. The aid agencies judged that these would have an inhibitory effect on tendering by British and American contractors and would give rise to high tender prices. As a result of these concerns a tripartite agreement was signed by the three governments in 1979. This exempted British and American contractors and consultants working on the project from the decennial liability provision of the Egyptian civil code.

15.　It was also decided that special construction insurance arrangements should be made. CWO provided comprehensive project insurance cover for contractors to the requirements of the Fédération Internationale des Ingénieurs-Conseils (FIDIC) Conditions of Contract and for consultants during the construction period. The insurance was placed on the London market by a public sector Egyptian insurance company with risks being reinsured worldwide. Also, the main consultants were required to provide professional indemnity

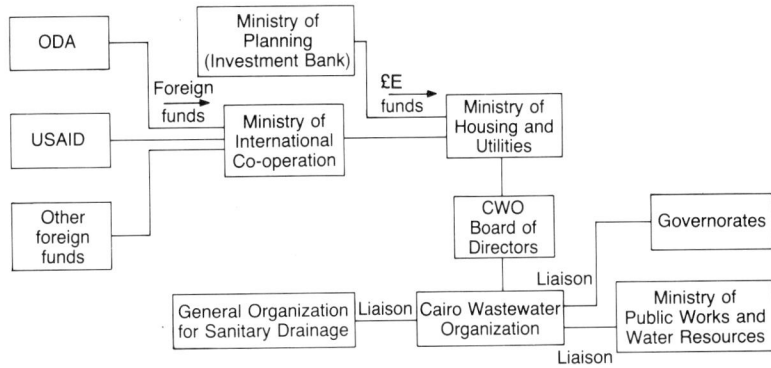

insurance for £10 million—this cover to be maintained for a period of ten years after the end of the project. The insurance is arranged through the UK Association of Consulting Engineers' scheme, with part of the risks being reinsured in the USA.

16.　At the time that the initial funding arrangements were being made it was recognized that there was insufficient sterling funding for the East Bank scheme. Through British Water and Wastewater Limited, who were responsible for bringing together many of the British interests involved, the GOE took advantage of an offer from the British merchant bank, Samuel Montagu (later Midland Montagu) for a commercial loan of £100 million backed by the UK Export Credit Guarantee Department (ECGD).

17.　The amounts provided by these three initial sources of foreign currency have been increased substantially to meet specific needs as the project developed. Recently major additional foreign currency funding has been provided by the Government of Italy (in 1991 for the construction of Gabal el Asfar Treatment Plant) and the European Investment Bank (in 1993 for the construction of the East Bank Branch Tunnels). The sources and amounts of funds currently committed to the project are shown in Table 1, amounting to an equivalent total of £1231 million at March 1993 exchange rates.

Fig. 1. Government of Egypt's arrangements for project implementation

Table 1. Sources of funding

Source of funds	Amount committed: millions			
	£E	£ sterling	$	ECU
Government of Egypt	1193·85	26·94		
UK ODA grants		65·33		
UK bank loans		185·00		
USAID grants	347·64		733·88	
Italian grants/loans			96·57	
EIB loans				45·00
Total	1541·49	277·27	830·45	45·00
Total: £ sterling equivalent at March 1993 exchange rates	327·28	277·27	588·97	37·19

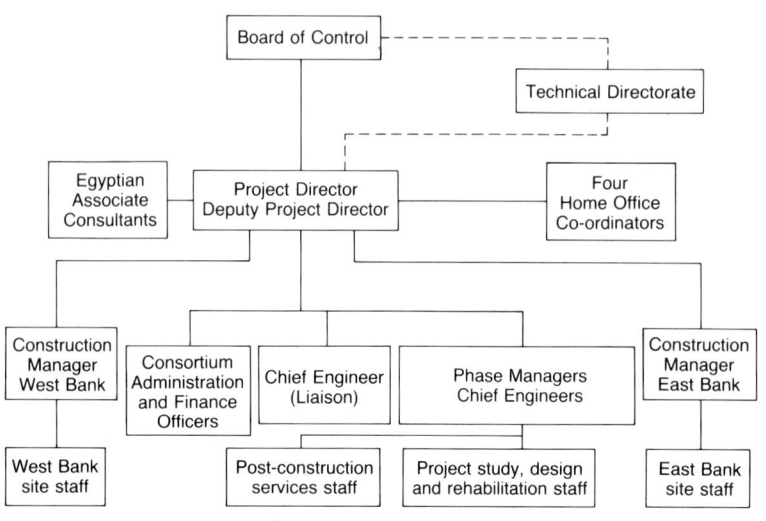

Board of Control

Technical Directorate

Egyptian
Associate
Consultants

Project Director
Deputy Project Director

Four
Home Office
Co-ordinators

Construction
Manager
West Bank

Consortium
Administration
and Finance
Officers

Chief Engineer
(Liaison)

Phase Managers
Chief Engineers

Construction
Manager
East Bank

West Bank
site staff

Post-construction
services staff

Project study, design
and rehabilitation staff

East Bank
site staff

Fig. 2. AMBRIC organization

Arrangements for consulting services

18. As a consequence of the initial funding arrangements it was agreed by the ODA and USAID that they would jointly provide technical assistance to the GOE. The ODA appointed Taylor Binnie and Partners (TBP) to continue for the next phases of the project and a joint venture of American consultants Black and Veatch International and Camp Dresser & McKee Inc. was selected following competition. The four British and American firms agreed to join together to form a composite joint venture under common management to be known as American British Consultants (AMBRIC), with each firm having comparable input and responsibility.

19. AMBRIC formed an association with local consultants as subconsultants. For the initial stages of the project (1979–1986) the selected local consultant was EGYCON, a consortium of three Egyptian firms (Atco, Sanes and Tencon). In 1986 the MOHU decided to terminate the involvement of EGYCON. Following this AMBRIC entered into an association with the firms of Arab Consulting Engineers (ACE) and Sabbour Associates. These arrangements have continued to date.

20. In addition to the principal consultants a number of specialist consultants have provided services as subconsultants to AMBRIC. In particular Mott Hay & Anderson (now Mott MacDonald) have participated throughout the project on the East Bank tunnelling work. To satisfy USAID regulations, a proportion of the American technical assistance has been provided by Minority Business Enterprises under subcontract to the American principal consultants.

21. In view of the special nature of the consultancy arrangements it was decided by the funding agencies to appoint AMBRIC as Programme Manager and Engineer for all phases of the work. A comprehensive agreement was drafted that set out the duties and obligations of the parties and provided a mechanism for the arrangement of work elements. Because it was recognized that the project would have a duration in excess of five years, the main agreement

contained only the detailed scope, staff resources and costs for four initial elements of work

(*a*) an inception report
(*b*) a project for rehabilitation of the existing sewerage system
(*c*) the training of GOSD staff
(*d*) the review and updating of the TBP master plan.

22. The services flowing out of the master plan review, involving detailed design and preparation of tender documents, tender appraisal, services during construction, training and operation assistance, were arranged by a mechanism known as work orders and amendments to work orders. Once a new element of work was defined the details of the work, the resources needed for its implementation, and the associated costs were negotiated by CWO and AMBRIC, and were reviewed and approved by the CWO Board of Directors and the appropriate funding agencies. During the course of the project about 70 such work orders and amendments have been arranged.

23. AMBRIC was set up and has operated throughout its existence as a fully integrated organization staffed by the four joint venture firms and its specialist subconsultants and supported by its local associates. The day-to-day direction of the project has been provided by a full-time resident Project Director (PD) assigned from one of the joint venture firms in rotation. The PD has been assisted by a deputy from a separate firm so that both the UK and the USA have always been represented at top management level.

24. The organization of AMBRIC supporting the management team has been kept flexible to suit the needs of the project during the various phases of the work. A typical structure is shown in Fig. 2. Throughout the duration of the project every effort has been made to ensure close co-ordination of the AMBRIC teams working on the East and West Bank schemes to provide a quality assurance check on work outputs, and to ensure as far as possible uniformity of design criteria and solutions.

25. The PD and his staff are responsible to a board of control (BOC) comprising a principal of each of the four joint venture firms. The BOC meets in Cairo at four-monthly intervals to review progress, to ensure that the team have adequate resources for the work in hand, and to discuss progress with the Client. The BOC is assisted by a Technical Directorate (TD) comprising senior technical representatives of the four firms or their specialist nominees.

Master plan review

26. Following the signing of the main agreement between CWO and AMBRIC in April 1979, AMBRIC's first task was to review the 1978 master plan. In essence 16 strategic discharge options were considered, based on 11 possible treatment plant sites within the project area.

27. The options for discharge points included various locations on the Mediterranean Sea, the Red Sea, the Great Bitter Lake, the Nile upstream of the Nile Barrage north of Cairo, the Rosetta branch of the Nile, agricul-

tural drains and the desert. Main collector and conveyance proposals were also reviewed and further comparisons and sensitivity analyses conducted on two alternative tunnelling schemes for the East Bank main collector: a relatively shallow tunnel approximately 20–26 m deep through soft alluvial deposits or a rock tunnel at a depth exceeding 150 m. Environmental aspects were considered briefly in the review, and USAID subsequently arranged an environmental assessment of the West Bank scheme which was carried out by a separate consultant.

28. The comprehensive review commenced in December 1979 and was submitted to CWO in the form of the Design Inception Report in January 1982. The recommendations were adopted by CWO and they formed the basis for the detailed design undertaken through subsequent work orders. These recommendations differed very little from the earlier TBP master plan. The finalized schemes on both the East and West Banks of the Nile are shown in Fig. 3.

Outline of the project works

East Bank scheme

29. The main works on the East Bank of the Nile comprise the following main elements.

(a) A 12·1 km long deep spine tunnel of 4 and 5 m diameter, with 35 km of interconnecting branch tunnels, flowing in a north-easterly direction to the Ameria Pumping Station Complex. The tunnels pass along the routes of busy streets at depths of up to 26 m below the existing sewer and Metro systems and below the groundwater table. The majority of existing sewers can therefore drain by gravity into the new main tunnel collector.

(b) Two new pumping stations built within the existing Ameria Complex. One, of 32 m depth and with a pumping capacity of $2·2 \times 10^6$ m³/day, serves the spine tunnel, and the other, a screw pumping station of $0·56 \times 10^6$ m³/day capacity, serves the existing high-level collectors already

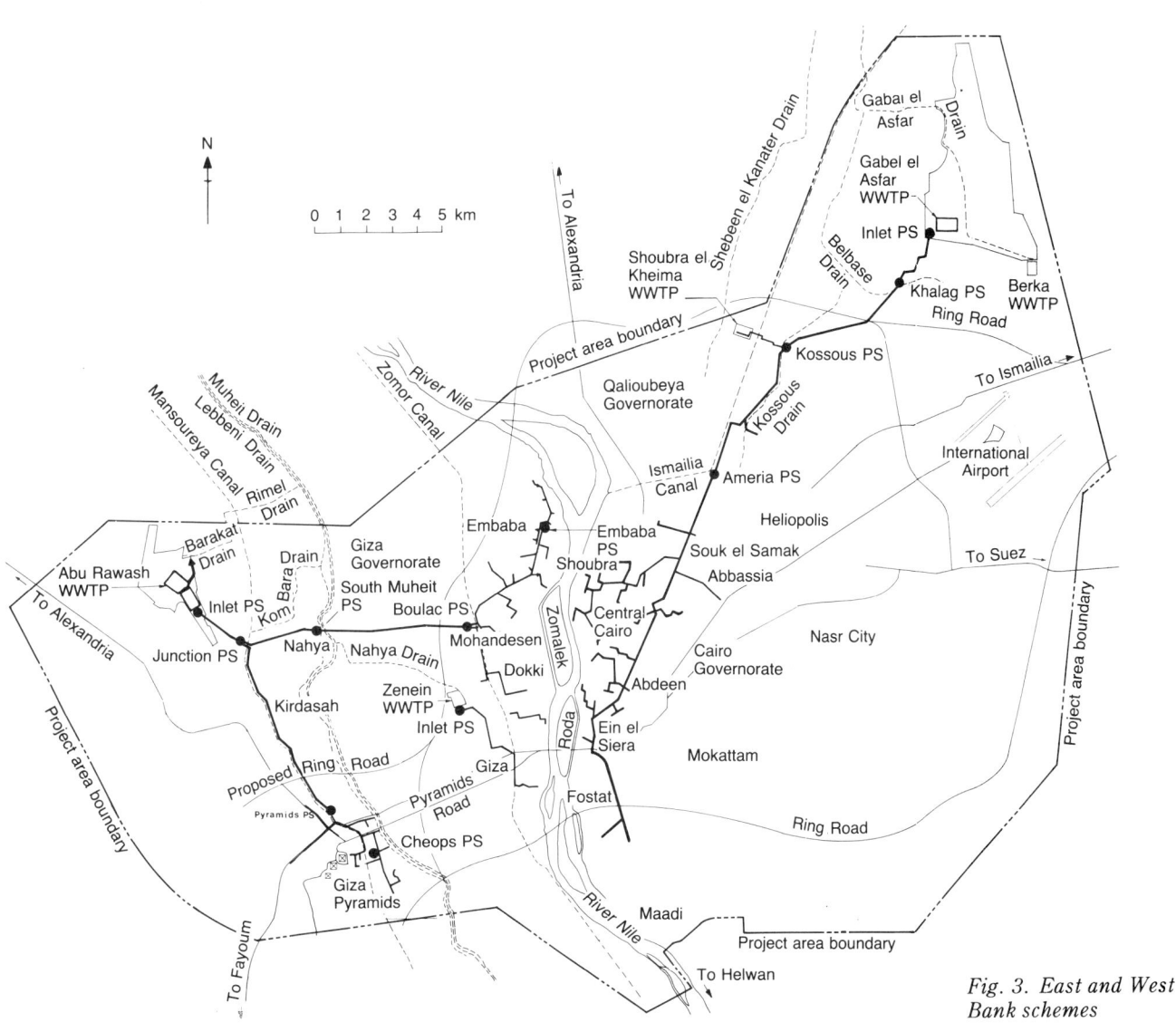

Fig. 3. East and West Bank schemes

on the site. The complex is now able to collect and discharge flows of up to 1.85×10^6 m³/day north-east via the new conveyance systems to Gabal el Asfar and Shoubra el Kheima treatment plants (the latter of 0.6×10^6 m³/day capacity under construction by GOSD) or through existing pumping mains flowing east to the old emergency canals and a new treatment plant at Berka (also constructed by GOSD with a capacity of 0.6×10^6 m³/day to primary standard at present).

(c) A new 14 km conveyance system from Ameria to Gabal el Asfar which allows wastewater to be carried through large multi-barrelled box culverts and major intermediate screw lift pumping stations at Kossous and Khalag to well beyond the urban boundary.

(d) A new 1×10^6 m³/day secondary treatment plant at Gabal el Asfar, which by 1995 will enable all wastewater collected on the East Bank to be treated prior to discharge to the existing Belbeis drain which flows north 159 km to Lake Manzala on the Mediterranean coast.

Specific details on the construction of these major elements are provided in references 2–4.

West Bank scheme
30. The works on the West Bank are described in reference 5 and comprise six main elements

(a) a new Giza relief collector sewer pipe constructed to the existing 0.33×10^6 m³/day secondary treatment plant at Zenein which has been completely rehabilitated

(b) a new north-west collector from Embaba to Abu Rawash using multi-barrel culverts and pipes and with five in-line screw pumping stations

(c) a western collector serving the Pyramids area built to connect into the new north-west collector at Junction Pumping Station

(d) a new treatment plant at Abu Rawash, initially treating 0.4×10^6 m³/day to primary standards; the treated effluent eventually discharges to the Rosetta branch of the Nile via the existing agricultural drainage system.

(e) sewage sludge from the Zenein and Abu Rawash plants will be thickened and pumped west from Abu Rawash a distance of 30 km to disposal in large desert lagoons where the sludge will dry naturally; drying beds have also been built at Abu Rawash to provide some local drying of sludge for agricultural reuse as a fertilizer

(f) although not anticipated in 1982, an extensive programme of property connections and lateral sewer systems covering an area of 1500 ha in the unsewered Pyramids, Zenein and Embaba areas is also under way in a series of 54 so-called FAR (fixed amount reimbursable) contracts, and is being carried out by Egyptian public and private sector contractors, with about half being carried out by each sector.

Implementation method
Design development
31. In accordance with the established policy, both preliminary and detailed design work for the conveyance systems on both East and West Bank schemes were carried out in Cairo by integrated teams of British, American and Egyptian engineers. In some specific cases, such as the design of the major pumping stations and treatment plants, teams were set up in the home offices but close liaison was maintained with the resident team.

Contract strategy and packaging of East Bank contracts
32. In October 1981 CWO directed AMBRIC to prepare a 'contract strategy' for the first section of work under design, the East Bank scheme. The strategy was to set out the basis for the implementation of the work and for the preparation of contract documents. A similar strategy was later adopted for the West Bank scheme. Principal matters to be considered were

(a) the requirements of foreign and local currency and the scope for adjusting the relative proportions

(b) the packaging of contracts and the sources of funds for the various contracts

(c) a programme for implementation

(d) the type of contract and the form of tender documents

(e) procedures for the selection of contractors

(f) project management procedures.

33. The strategy took into account the constraints and requirements of CWO and other Egyptian authorities, the various funding agencies, funding limitations and other factors both physical and administrative. AMBRIC were assisted in the preparation of the contract strategy by project management specialists Martin Barnes & Partners.

34. The East Bank scheme is linear in nature, covers urban and rural areas, and involves many different types of plant and construction, which dictated the need for arranging the works into suitable contract packages.

35. The packaging and programming of the works was carried out as an iterative process. The first step was to subdivide the scheme into elements and to assign them a priority based solely upon public health and environmental considerations. This preliminary ranking gave priority to the relief of the overloaded sewerage system in the urban areas and the conveyance of the sewage to the urban boundary. This initial ranking was then reviewed and amended and a preliminary programme and cash-flow budget prepared. Factors taken into account included

(a) the priority preferences of CWO and other Egyptian authorities

(b) the likely impact of the project on local material resources, and the capacity of the local construction industry in both manpower and equipment

(c) the nature of the work and the need for participation of foreign (UK) contractors

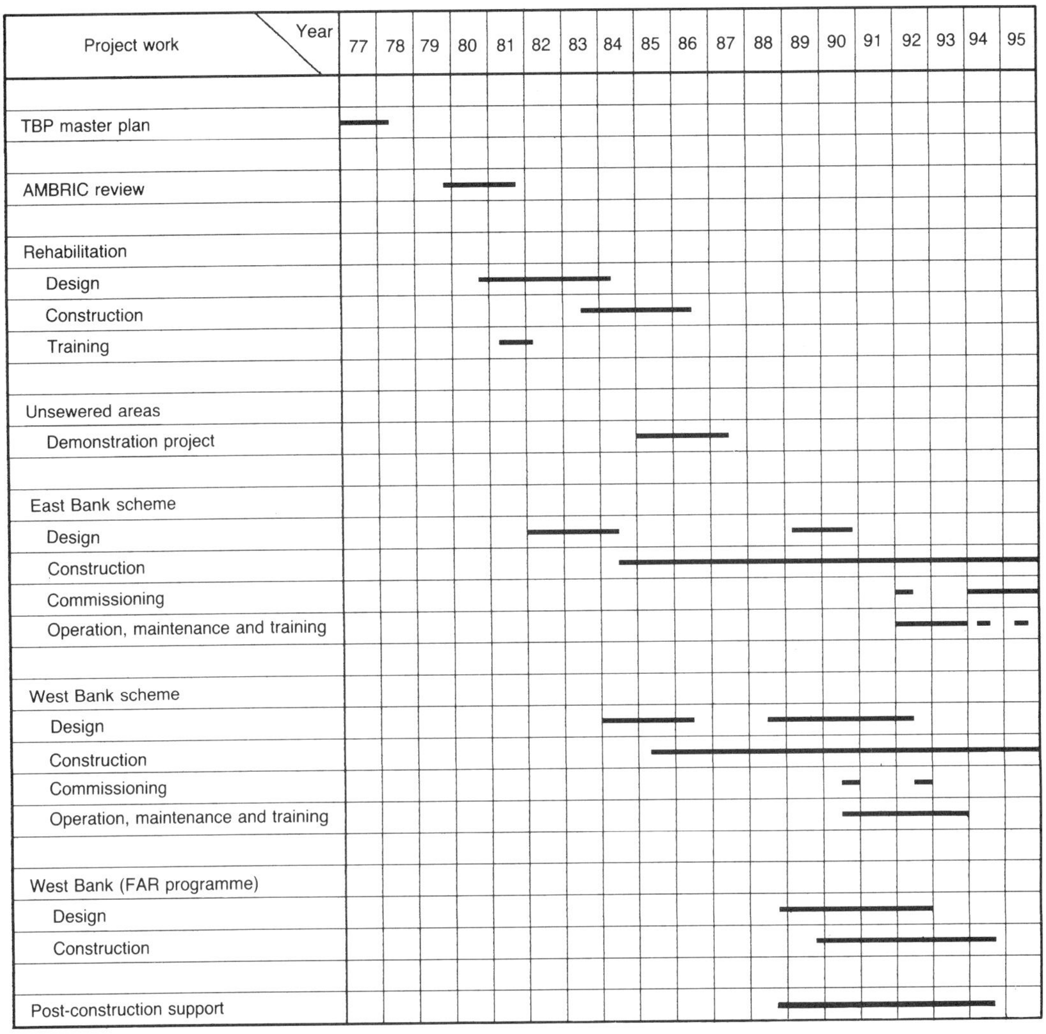

Project work \ Year	77	78	79	80	81	82	83	84	85	86	87	88	89	90	91	92	93	94	95
TBP master plan	▬	▬																	
AMBRIC review				▬	▬														
Rehabilitation																			
Design					▬	▬	▬												
Construction							▬	▬	▬										
Training					▬														
Unsewered areas																			
Demonstration project								▬	▬										
East Bank scheme																			
Design						▬	▬						▬	▬					
Construction								▬	▬	▬	▬	▬	▬	▬	▬	▬	▬	▬	▬
Commissioning															▬		▬		
Operation, maintenance and training																▬	▬	▪	▪
West Bank scheme																			
Design							▬	▬					▬	▬	▬				
Construction									▬	▬	▬	▬	▬	▬	▬	▬	▬	▬	▬
Commissioning													▪		▪				
Operation, maintenance and training													▬	▬	▬	▬	▬		
West Bank (FAR programme)																			
Design													▬	▬	▬				
Construction														▬	▬	▬	▬		
Post-construction support									▬	▬	▬	▬	▬	▬	▬	▬			

Fig. 4. Programme of completed, current and planned work

(d) the total funds available and the rate at which they would be provided

(e) the sources of funds and the requirements of the funding agencies.

36. The need for action within the urban area was urgent and widespread, and it was for this reason that CWO gave highest priority of all to rehabilitation of the existing system, as this would provide the earliest benefits.

37. Highest priority for new construction was given to the spine tunnel from Ein el Siera to Ameria, the Ameria complex and the first length of culvert from Ameria to the urban boundary. Second priority was given to extension of the conveyance system to Gabal el Asfar and to the construction there of the first stage of the wastewater treatment plant. Third priority was given to the extension of the spine tunnel to Maadi, the construction of branch tunnels and further stages of development of the Gabal el Asfar treatment plant.

38. All contracts have been based on the FIDIC Conditions of Contract, 3rd edition, with some modifications and additions. On the main West Bank contracts, works were to be carried out by American contractors. On the East Bank the tunnel and pumping stations were generally awarded to UK–Egyptian joint ventures, while the culvert contracts were awarded to local public and private sector contractors with UK management assistance. A general programme

of completed and planned works is shown in Fig. 4.

Rehabilitation of the existing sewerage systems

39. In order to gain time and early benefits, the first contracts to be implemented were for a US$95 million and LE38 million rehabilitation programme. More than 100 pumping stations and ejectors were retrofitted with new mechanical and electrical equipment, improved control systems, and improved operation and maintenance features. Six new pumping stations and some 47 km of sewers and force mains were constructed to increase capacity, to redirect discharges from surcharged sewers, and to replace existing pumping stations and force mains that were in extremely poor condition.

40. The pumping station rehabilitation effort was a significant success. It provided greater capacity and reliability of pumping capability, and a major reduction in the incidence of wastewater flooding in the streets of Cairo.

Construction

41. Construction of the priority contracts of the East Bank scheme started in 1984 and on the West Bank a year later. On the East Bank scheme 12 major contracts were awarded to foreign–Egyptian joint venture contractors with individual values of between £3 million

Table 2. Approximate material and construction figures

Project works	Tunnels constructed	21 km
	Culverts constructed	32 km
	Sewers and force mains laid	464 km
	Property connections made	150 000
	Pumping stations (constructed)	13
	Pumping stations (rehabilitated)	104
	Treatment plants (constructed)	2
	Treatment plants (rehabilitated)	1
Materials	Concrete	$2 \cdot 2 \times 10^6$ m^3
	Reinforcement steel	190 000 t
	Acid-resistant bricks	36×10^6

and £100 million. On the West Bank 11 major contracts were awarded to American firms with values between US$9 million and US$115 million and a further 54 smaller contracts for secondary sewerage works to local public and private sector contractors each with a value of up to £E10 million.

42. The magnitude of the construction involved in the project to date may be gauged by the approximate material and construction figures shown in Table 2. Summary programmes of the East and West Bank construction are given in Figs 5 and 6.

Commissioning of the new systems

43. Although the 26 km long spine tunnel and culvert system to Gabal el Asfar was almost complete by mid 1991 the major treatment plant contracts, awarded to Italian–Egyptian joint venture contractors, had not yet started. The Egyptian authorities perceived that there could be substantial public health and operational advantages in proceeding with partial commissioning of the scheme in advance of completion of the treatment works scheduled for 1995. It was therefore decided that wastewater be diverted into the spine tunnel system and pumped through Ameria as far as the Kossous pumping station, where a flow of up to $0 \cdot 65 \times 10^6$ m^3/day would be discharged into the existing and already highly polluted Kossous drain. The major difficulty in achieving this flow regime lay in clearing and increasing the capacity of existing drains from Kossous for a distance of 40 km downstream to enable them to accept the interim discharges and later the full phase 1 flows from the Gabal el Asfar and Berka treatment plants. These necessary works were not the direct responsibility of CWO but were carried out by the MPWWR.

44. The three main tunnel contracts had included provision for the connection of existing sewers through temporary grit traps and vortex drop structures. In view of revised plans for commissioning, these works were deleted from these contracts and combined in a separate Flow Diversion Contract. With a January 1992 date set for the ceremony to commemorate the operation of the new East Bank system a final 80 day programme of intensive activity started. In addition to finalizing completion of outstanding works at the Ameria and Kossous pumping stations, technical details of

the planned phased connection programme to the tunnel and Ameria complex pipework were worked out, drain-cleaning operations were intensified, and an initial operations and maintenance contract was finalized.

45. On 16 January 1992, the first line connections were made to the tunnel. By mid 1993, the flows into the tunnel averaged $1 \cdot 2 \times 10^6$ m^3/day, including a volume of $0 \cdot 1 \times 10^6$ m^3/day previously pumped from the East to the West Bank for disposal. The effect on the existing sewerage system has been significant. There has been a dramatic reduction in water levels in the system and in the incidence of sewage flooding in urban areas reported by GOSD. The relief of overloading within existing sewers at last permits a structured programme of sewer cleaning and rehabilitation not previously possible. As further branch tunnels are completed and connected so will more of the old ejector and district pumping stations be decommissioned—a total of 75.

46. On the West Bank, Abu Rawash treatment plant and the new northern conveyance system from the densely populated Embaba area were successfully commissioned in October 1992.

Operations and training

47. In anticipation of the operations phase of the project the staff of GOSD, the government authority responsible for future operations and maintenance of existing and new systems, were given preliminary training by AMBRIC in such aspects as pumping station and treatment plant maintenance. This training was complemented with hands-on site training by contractors on both the East and West Bank schemes. The general policy set by CWO has been to retain those expatriate contractors who provided and installed mechanical and electrical equipment for a further 2–3 year period after commissioning to prove the working reliability of the new facilities and to help train GOSD staff to take over and operate unassisted.

Programme and financial control

48. The control of programme and related cash flow was particularly important in view of the interdependence of the contracts and the numerous sources of finance. Monitoring systems for the programme and for management accounting were developed by AMBRIC based on MAPPS project monitoring software. Information on each running contract provided monthly by the contractors and resident engineers was processed and integrated to determine the overall project situation. The system also provided estimates of cash flow for the project as a whole and allowed the monitoring of key resources such as cement and bricks. This facility was particularly useful during the period of peak construction activity, when the availability of certain resources, in particular blue bricks forming secondary lining, caused constraints on progress and the need to establish project priorities.

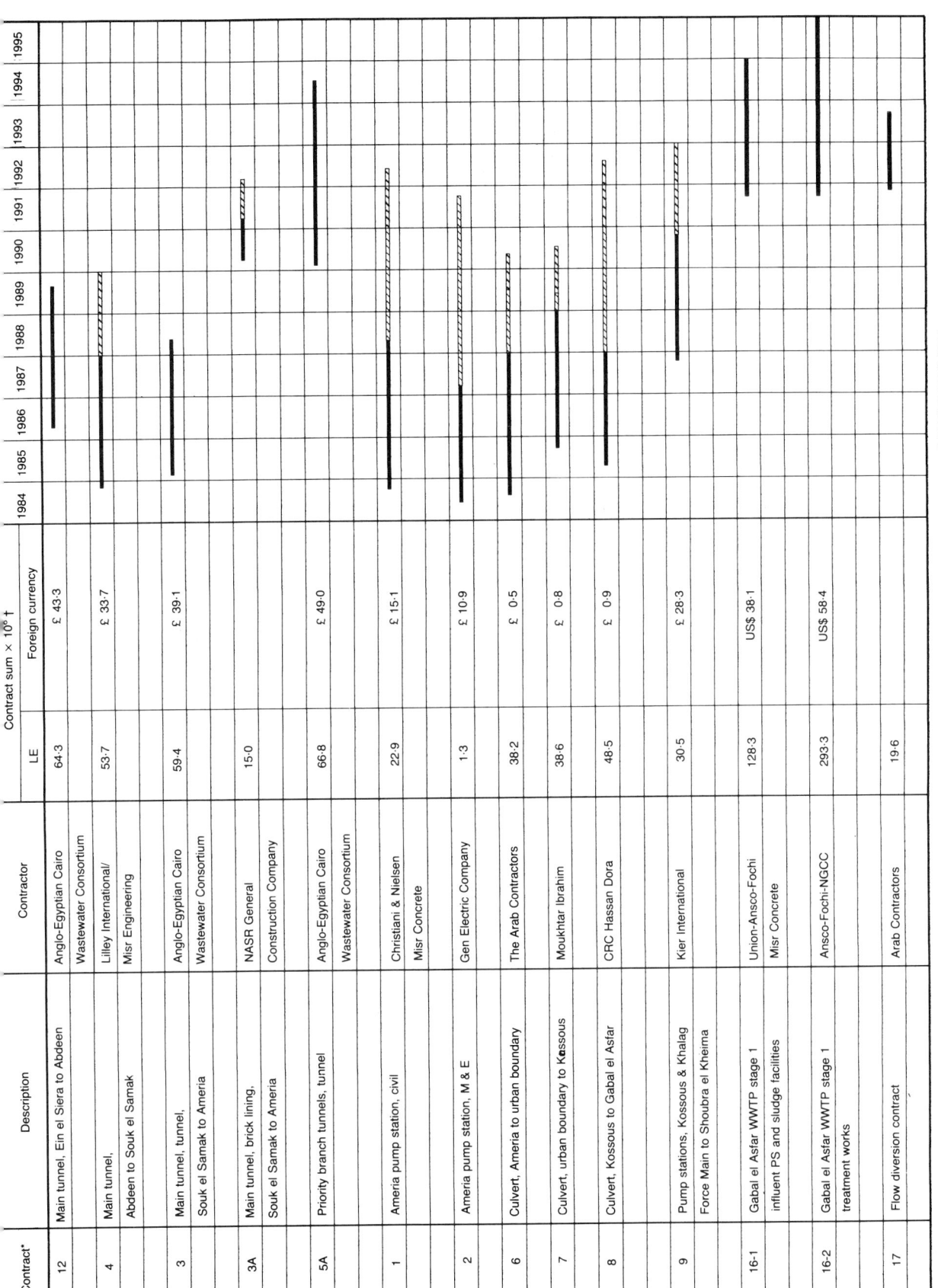

Contract*	Description	Contractor	Contract sum × 10⁶ †		1984	1985	1986	1987	1988	1989	1990	1991	1992	1993	1994	1995
			LE	Foreign currency												
12	Main tunnel, Ein el Siera to Abdeen	Anglo-Egyptian Cairo Wastewater Consortium	64·3	£ 43·3												
4	Main tunnel, Abdeen to Souk el Samak	Lilley International/ Misr Engineering	53·7	£ 33·7												
3	Main tunnel, tunnel, Souk el Samak to Ameria	Anglo-Egyptian Cairo Wastewater Consortium	59·4	£ 39·1												
3A	Main tunnel, brick lining, Souk el Samak to Ameria	NASR General Construction Company	15·0													
5A	Priority branch tunnels, tunnel	Anglo-Egyptian Cairo Wastewater Consortium	66·8	£ 49·0												
1	Ameria pump station, civil	Christiani & Nielsen Misr Concrete	22·9	£ 15·1												
2	Ameria pump station, M & E	Gen Electric Company	1·3	£ 10·9												
6	Culvert, Ameria to urban boundary	The Arab Contractors	38·2	£ 0·5												
7	Culvert, urban boundary to Kossous	Moukhtar Ibrahim	38·6	£ 0·8												
8	Culvert, Kossous to Gabal el Asfar	CRC Hassan Dora	48·5	£ 0·9												
9	Pump stations, Kossous & Khalag Force Main to Shoubra el Kheima	Kier International	30·5	£ 28·3												
16-1	Gabal el Asfar WWTP stage 1 influent PS and sludge facilities	Union-Ansco-Fochi Misr Concrete	128·3	US$ 38·1												
16-2	Gabal el Asfar WWTP stage 1 treatment works	Ansco-Fochi-NGCC	293·3	US$ 58·4												
17	Flow diversion contract	Arab Contractors	19·6													

Original contract dates

Extended contract dates

*Contracts listed from upstream to downstream

† Amounts are in LE and foreign currency components at time of award (LE4·71 = PDS1, March 1993)

Fig. 5. East Bank construction programme

Contract	Description	Contractor	Contract sum × 10⁶ * LE	US$
20A	Embaba sewers and collectors	Harbert/Jones		115·0
21	Culverts, Boulac to Abu Rawash	Fru - Con		31·3
22	Pumping stations	Fru - Con		43·6
23	Zenein pumping station	Sadelmi NY		14·4
23A	Collector and Giza relief system	SEDE	21·0	
25	Pyramids culvert	AICI		29·7
26	Pyramids pumping station	Fru - Con		9·5
27	Pyramids collector	Harbert Jones		
28	Cheops pumping station	Morrison Knudsen Corp.		
29	Abu Rawash WWTP	Sadelmi USA		114·0
30	Abu Rawash effluent disposal system	ARABCO	41·0	—
31	Zenein WWTP rehabilitation	Sadelmi NY	5·0	69·8
33A	Western Desert sludge pipeline and disposal	Harbert Jones		24·4
33B	Western Desert sludge pump station	Caddel Corp.		7·8
35	Sludge pilot facilities	Under tender		4·2 (est.)

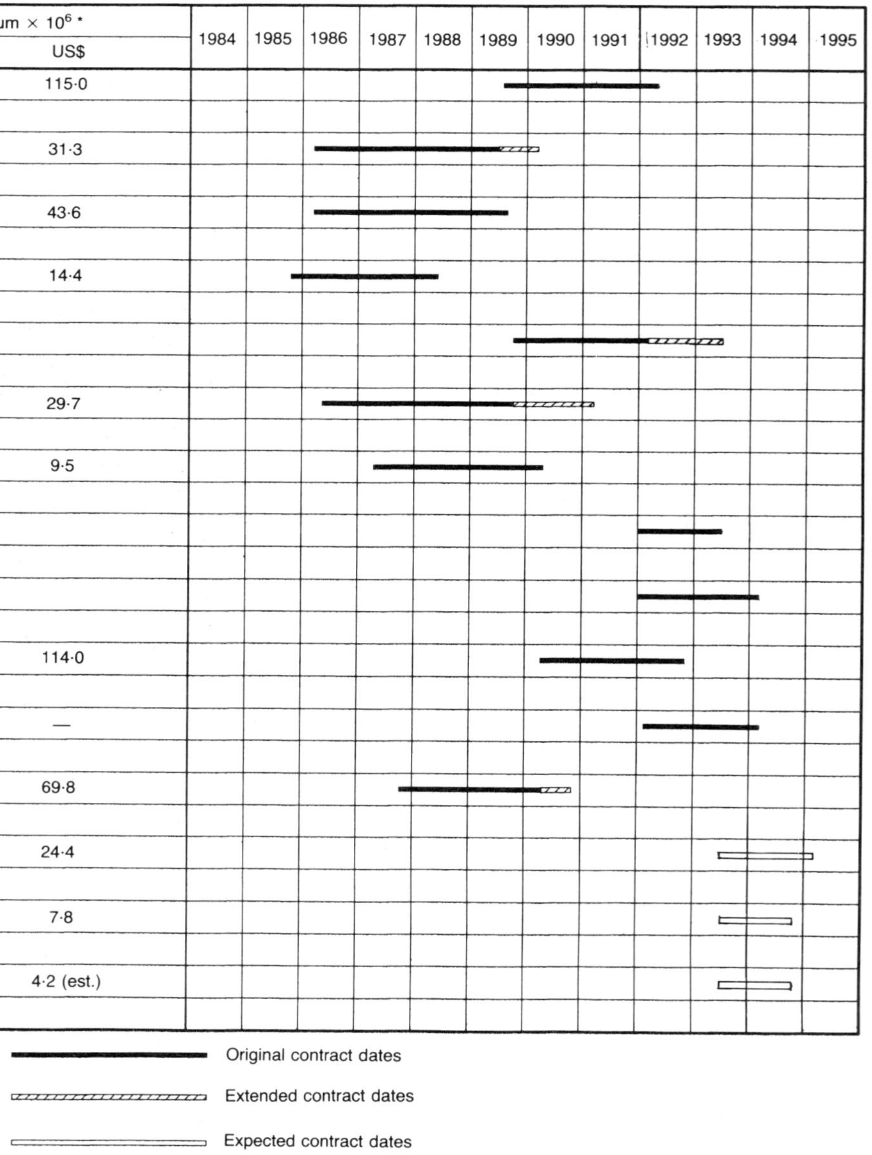

Years shown: 1984, 1985, 1986, 1987, 1988, 1989, 1990, 1991, 1992, 1993, 1994, 1995

———— Original contract dates
///// Extended contract dates
▭▭▭ Expected contract dates

* Amounts are in LE and US$ components at time of award (LE3·33 = US$1, March 1993)

Fig. 6. West Bank construction programme

Public communications and education

49. Prior to the new facilities being put into use, the MOHU were anxious to provide the public with more information about the project. In July 1991 AMBRIC presented a public communications work plan to provide a structured programme of information. This included an educational programme through the media and in schools on the use and misuse of the new sewerage systems in Cairo.

Records management

50. During the course of the project period it is estimated that more than 5 million pieces of paper have been generated. Of this amount, 20 000 drawings and more than 300 000 other documents have been microfilmed, the remainder being archived or destroyed.

Summary of progress

51. Given the complexity of the works, there have been relatively few construction related delays. The two most significant of these were the caisson sinking and tunnel freeze difficulties experienced at the Ameria Tunnel Pumping Station, discussed in reference 4, and a long-running problem in timely delivery of adequate quantities of the clay bricks which formed the internal acid-resistant linings to tunnels and culverts.

52. In total, 36 million blue bricks were required for the project, manufactured from Aswan clay to Egyptian Standard 41 by local public and private sector brick suppliers. Supply of bricks to the contractors never met scheduled requirements. Contractual problems resulted, and in 1988, when deliveries had almost ceased, AMBRIC were required to take steps to ensure that adequate supplies were directed to priority contracts in order to mitigate delays and additional costs to the project.

53. At the time of preparation of the East Bank contract strategy in 1981, it was estimated that the foreign currency funding available would be sufficient to construct the main tunnels and conveyance system, including the pumping stations, and a first phase of the Gabal el Asfar treatment plant with a capacity of 0.25×10^6 m^3/day. In the event the funds were not sufficient for the Gabal el Asfar works, which were taken out of the UK funding package. The available remaining UK funds were reallocated to enable the priority branch tunnel contract 12/5A to be constructed as part of the first phase works. Additional funds from the European Investment Bank have been arranged to construct a further major package of branch tunnels, and Italian government funding agreed in 1990 has enabled contracts to be awarded for construction of the full 1×10^6 m^3/day plant at Gabal el Asfar.

54. The overall programme and objectives of the East and West Bank schemes in meeting the city's needs are now well advanced with many of the major elements of the systems in operation. The institutional aspects which will ensure long-term operational sustainability of the new facilities are being addressed by the Egyptian government with support from the international community.

References

1. ROBERTS D. G. M., SALEM S. A. and FLAXMAN E. W. Introduction and historical overview. *Proc. Instn Civ. Engrs, Civ. Engng, Greater Cairo Wastewater Project*, 1993, 3–7.
2. FLINT G. R. *et al.* Collection system: tunnels. *Proc. Instn Civ. Engrs, Civ. Engng, Greater Cairo Wastewater Project*, 1993, 18–33.
3. GRIMES J. F. *et al.* Conveyance system: pumping stations and culverts. *Proc. Instn Civ. Engrs, Civ. Engng, Greater Cairo Wastewater Project*, 1993, 34–47.
4. TAYLOR R., HALIM M. and WISHART M. Gabal el Asfar treatment plant. *Proc. Instn Civ. Engrs, Civ. Engng, Greater Cairo Wastewater Project*, 1993, 48–55.
5. MILLER G. R. and KACHINSKY R. West Bank scheme. *Proc. Instn Civ. Engrs, Civ. Engng, Greater Cairo Wastewater Project*, 1993, 56–59.

Proc. Instn
Civ Engrs,
Civ. Engng,
Greater Cairo
Wastewater
Project,
1993, 18–33

Paper 10229

Collection system: tunnels

G. R. Flint, BSc, ACGI, MICE, I. H. Elliott, BSc, FICE, W. Foreman, BA, MICE,
B. M. Griffin, BSc, ARSM, T. McDonald, BSc, FICE, and R. J. Reed, FICE

G. R. Flint,
Acer Consultants
Limited;
Senior Resident
Engineer,
AMBRIC

I. H. Elliott,
Managing Director,
Mott MacDonald
—Civil; Chief
Engineer Design,
AMBRIC
(1982–84)

W. Foreman,
Formerly
Binnie & Partners;
now Mott MacDonald
Group; Resident
Engineer, AMBRIC
(1985–88)

B. M. Griffin,
Edmund Nuttall Ltd;
Construction Manager,
CWC

T. McDonald,
Director,
Balfour Beatty
Construction
International;
Project Manager,
CWC (1985–89)

R. J. Reed,
General Manager,
Charcon Tunnels
Ltd;
Project Director,
Lilley–Misr Eng
J.V. (1984–90)

■ Introduction

This Paper describes some of the planning and design considerations of the tunnel system which leads to Ameria Pumping Station.[1] It then describes the construction work carried out under the various contracts highlighting the method of approach adopted by the different contractors and the problems encountered during construction.

2. The East Bank Scheme is designed to serve a population of over 8 million within an area of about 130 km². Wastewater flows were calculated from a detailed land use and population assessment followed by an estimation of the corresponding drainage requirement. Flows for each district (*kism*) were assessed from existing flow records, field measurements and estimations based on water consumption data. Estimated daily average wastewater flows were thus derived for each *kism* and drainage area for the years 1980, 1985, 1990, 2000 and for the planning horizon.

3. The new deep level gravity system is located below the existing sewerage system at depths of between 12 and 22 m below ground level. The main tunnel runs approximately north-east, along the route of the ancient Pharoanic Amnis Trajanus Canal, to the new deep level pumping station at Ameria. A series of smaller branch tunnels feed into the main tunnel at intervals making an overall system of about 48 km (see Fig. 1). The main tunnel is of 4 m finished internal diameter for the upstream 6·6 km and increases to 5 m for the remaining 5·5 km. The branch tunnels are of 2·5, 1·8 and 1·5 m diameter. The drainage of the whole area is by gravity.

Design features

4. The system is designed in accordance with the Ackers White equations for sediment transport to provide at least 200 mg/l grit transport capacity under minimum flow conditions and to have a minimum mean velocity of 0·6 m/s at the lowest average predicted flow. The hydraulics were refined and finalized using a specially adapted computer programme. Using the above criteria, minimum gradients of 1 in 1090 were established for the largest tunnels of 5 m and 1 in 950 for the smallest 1·2 m tunnels. For comparison, the first main collector built by the year 1913 with an internal diameter of 1·6 m has a gradient of 1 in 2500.

5. A study of wastewater strength and composition showed that the BODs recorded were consistent with a wastewater of above average strength, principally of domestic origin. An assessment of corrosion potential indicated that severe corrosion of the system infrastructure might occur within a few years of commencing operation. Consequently, tunnels and shafts are provided with a corrosion resistant lining of

blue engineering bricks with chemical resistant resin mortar pointing, polyvinal chloride (PVC) membrane soffits in shafts and special benching screeds. Entry from the existing system into the tunnel system has generally been provided by a vortex drop shaft with tunnel junctions entering at an acute angle. Model tests were carried out for both these structures to investigate the overall hydraulic performance, quantify energy losses, observe sediment deposition and minimize turbulence and eddy effects.

6. It was essential to minimize settlement effects in the dense urban environment of Cairo, where many of the buildings are old and on relatively shallow pad or raft foundations. Wherever possible, the tunnels are routed under streets to minimize property purchase or occupations, legal and other such administrative problems, and provides working areas. This results in many of the tunnel drives being curved. The minimum designed radii to cope with such constraints varied from 400 m to 200 m for the 5·0 m dia. and 1·8 m dia. tunnelled sewers respectively.

7. Access shafts are provided at junctions and at places where changes of direction, diameter and gradient occur. Vertical connections were generally made by a vortex drop shaft. Many of the shafts were used for construction. Wherever possible, shaft diameters were standardized at approximately 8·0 m dia., except where changes in direction and incoming branch connections necessitated increases up to a maximum of 10·0 m dia.

8. Site investigations revealed that the new tunnelled sewers would be well below groundwater level, and located mostly in fine, medium and coarse sands containing layers and lenses of clay or silt with occasional gravel and cobble bands, except the southerly tunnel under Fostat which would be in weathered, weak to moderately strong fissured limestone. A geological long-section along the main spine tunnel route is shown in Fig. 2. The combination of granular soils and a high groundwater table creates extremely difficult shaft sinking and tunnelling conditions, requiring special design considerations.

Advance works

9. Prior to work commencing, dilapidation surveys were undertaken of all properties and structures within the predicted shaft sinking and tunnelling zones of influence. A comprehensive system of ground settlement monitoring was instigated. Survey stations were established at predetermined distances along the whole route and were monitored throughout the tunnelling operations. Tunnel driving was carried out with the minimum of surface disturbance and disruption and settlement was generally within the predicted range of up to

40 mm, depending on the sizes and depths of the tunnels.

10. Trial excavations and electronic surveys were carried out to determine the position of utilities and the need for diversions. A Utility Liaison Committee was established with representation from all the utility authorities. Foundation inspection pits were also undertaken to evaluate the type and condition of foundation and the bearing material. A precision survey grid of benchmarks and survey points was established throughout the project area with local detailed mapping surveys carried out at shaft location, etc. A traffic management team was formed to prepare temporary traffic schemes in agreement with the City Traffic Police and public transport operators.

Construction considerations

11. Shafts were designed to be sunk under compressed air to balance the groundwater pressure and prevent inflow of the granular soil. The standard design of shafts comprised an upper reinforced concrete circular structure sunk as a caisson, with a lower segmental section constructed by underpinning. Ground treatment and compressed air was specified for breaking out of and into the shafts, and for constructing adits for future incoming tunnelled branch connections.

12. The design for the primary lining to the main tunnels is of a bolted segmental form developed for use with slurry tunnelling machines. The segments are tapered to enable the negotiations of curves and plane correction without special packing between the joints. Segments are provided with grooves for syn-

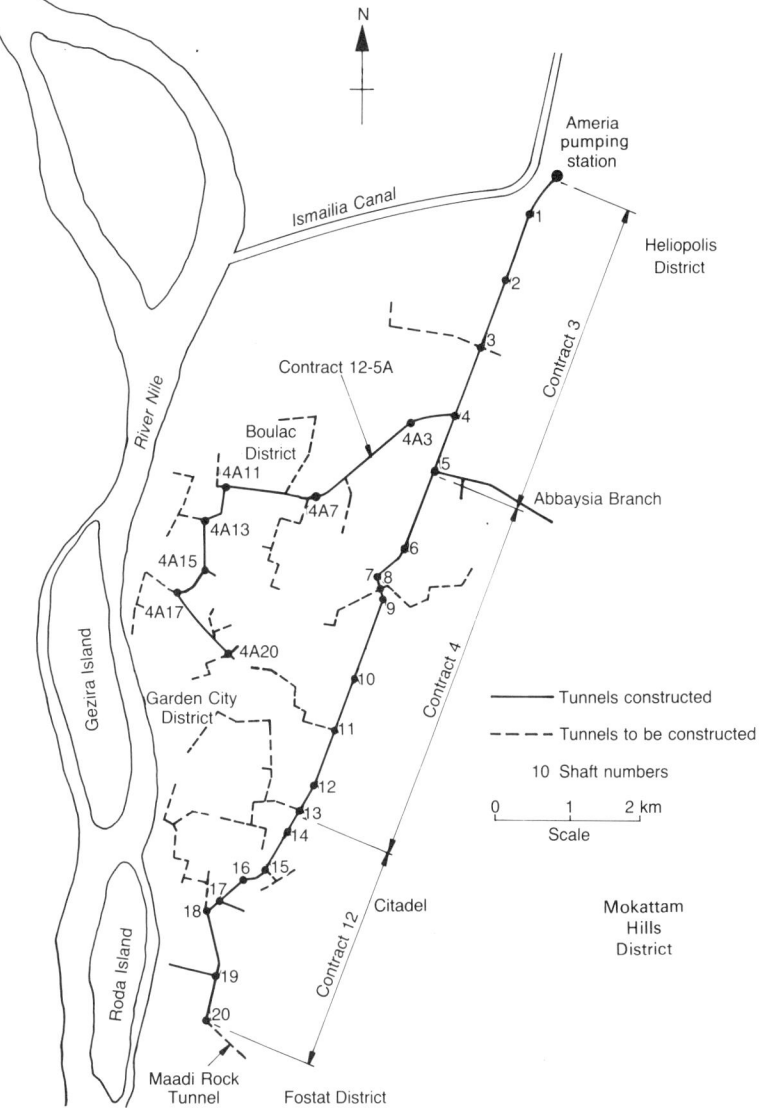

Fig. 1 (right). Present status of tunnels

Fig. 2 (below). Geological profile along tunnel route

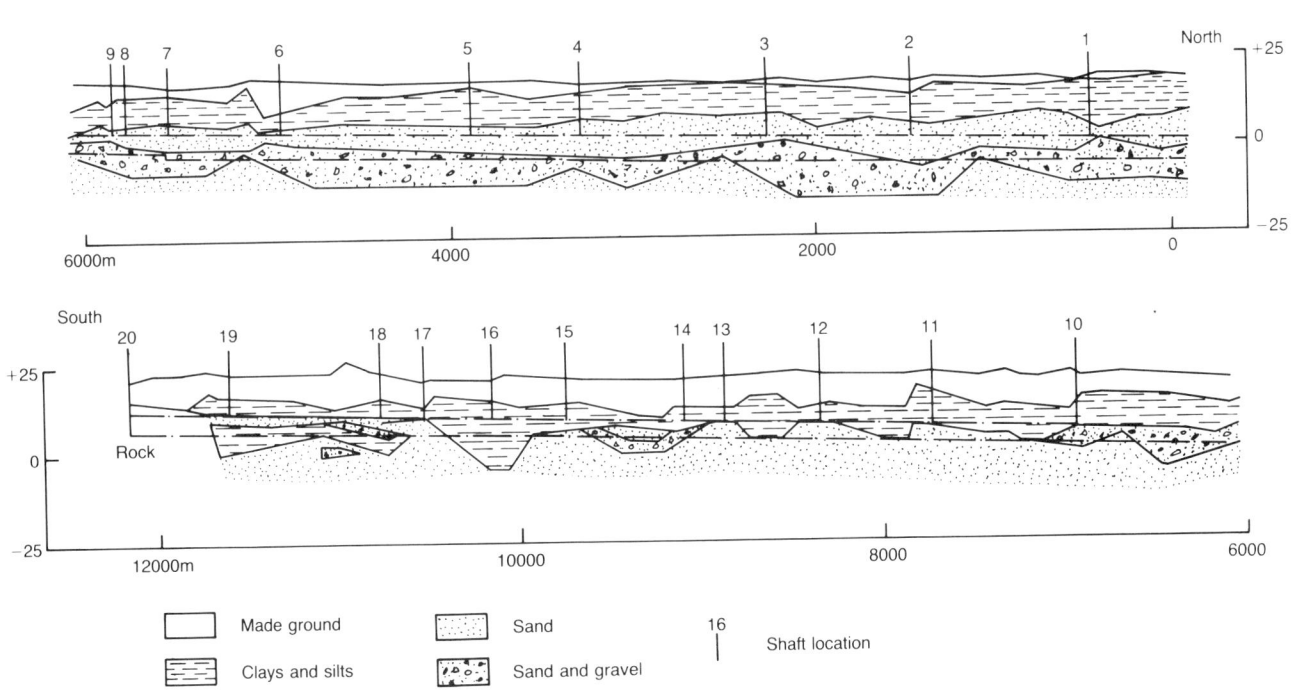

Table 1. Main features of tunnelling

Contract	Contractor	Length: km	Finished: m ID	Tunnel: m ID	Tunnel construction methods	Number of shaft	Shaft construction methodsds
3. Ameria to Souk El Samak	CWC	3·9	5·0	5·52	Bentonite tunnelling machines	17	Full depth caissons* sunk under compressed air
Abbasya Branch		1·6	2·5/1·8	2·87/2·2	Shields plus compressed air		
4. Souk El Samak to Abdeen	Lilley–Misr JV	1·6 3·3	5·0 4·0	5·52 4·57	Bentonite tunnelling machines	11	Part caisson/part underpinning sunk under compressed air
12. Abdeen to Ein El Siera (Fostat)	CWC	0·9	4·0	4·57	Bentonite tunnelling machines*	13	12 full depth caissons* sunk under compressed air and one in rock in 'free air'
		2·4/0·8	4·0/1·2	4·57/1·62	Shields plus compressed air		
12/5A. Boulac spine tunnels	CWC	2·5	2·5	2·87	Earth pressure balance machines†	21	Full depth caissons sunk under compressed air
		4·0	1·8	2·20			

* Contractor's alternative construction option
† Contractor's alternative tender offer

thetic rubber gaskets towards the extremity of their mating faces, to effect immediate sealing on erection. Where necessary they also incorporated caulking grooves on their inner edges for secondary sealing purposes. Key segments are wedge-shaped to allow insertion within the tailskin of the tunnelling machines without gasket damage. Segments for the smaller compressed air driven branch tunnels are of the standard parallel-sided panelled type with caulking grooves. The negotiation of curves was achieved by packing between the circumferential joints.

13. A number of different tunnelling methods was used successfully. Bentonite slurry machines were considered to be more suitable for the larger tunnel faces than compressed air with open-faced shields and machines, and the bentonite technique was specified for driving the majority of the spine tunnel. The smaller tunnels were designed to be driven in compressed air using hand or mechanical shields with the aid of ground treatment at the contractor's discretion. Earth pressure balance machines were used on the branch tunnels. Table 1 shows the main components on the four tunnel contracts.

Main tunnel (contracts 3 and 12)

14. The main spine tunnel between Ein el Siera and Ameria (see Fig. 1) was divided into three contracts (3, 4 and 12). Of these, the upstream section between Ein el Siera and Abdeen (contract 12) and the downstream section between Souk el Samak and Ameria (contract 3) were both won by a consortium of British and Egyptian contractors, the Cairo Wastewater Consortium (CWC, see box). Because of the different geotechnical conditions

of the two sections of tunnel, the Contractor needed to adopt different tunnelling methods.

Shafts: full depth caisson concept
15. Prior to starting work on contract 3 the Contractor requested approval to change the method of construction of the shafts from that envisaged at the design stage to a full depth caisson technique. The alternative method proposed sinking the whole shaft as a pneumatic monolithic reinforced concrete caisson with the tunnel eyes cast into the structure at the correct final levels (Fig. 3). This method was proposed because it was believed that it would be quicker to construct, significantly reduce the problems of compressed air loss compared with underpinning, and ultimately lead to less exposure of personnel to compressed air working.

16. This proposal was reviewed in the light of the need to restrict the inevitable escape of compressed air into the surrounding ground and the benefits the change might produce for the public at large. These included the reduced risk of adverse effects on the already high groundwater regime, such as flooding of basements, soil migration and public concern over the visual effects of air escapes. Consent was granted provided the Contractor took responsibility for the design of the caissons and had them verified by an independent consultant, approved by the Engineer, and took the risk of the full-depth caissons failing to go down and then having to revert to underpinning.

17. In the event it was found that once the full reinforced concrete shaft structure had been constructed above ground it could be sunk as a caisson within 7 to 10 days, and also that the quantity and pressure of compressed air

Cairo Wastewater Consortium
Tarmac Overseas Ltd (UK)
Balfour Beatty Construction Ltd (UK)
Cementation International Ltd (UK)
Edmund Nuttall Ltd (UK)
The Arab Contractors (ARE)

could be controlled to within fine limits. There
was minimal escape of air under the cutting
edge and no adverse effects on the surrounding
ground. No shafts failed to go down and subsequently the Contractor adopted this full-depth
caisson technique for contract 12. In all
29 shafts, ranging in internal diameter from
4·85 m to 10 m, were constructed by this
method with a maximum pressure of 2·1 bar in
the working chamber.

Shaft design and construction

18.   Calculations showed that the caissons
would need kentledge to overcome skin friction
and uplift from the compressed air. The excavated sand was placed on the air deck which
had been designed for the purpose. Resistance
to sinking was generally greater than anticipated and more sand kentledge than calculated
was used in most cases.

19.   Small diameter (75 mm) pipework was
cast into the caisson walls to allow bentonite
injection into the annulus around the caisson
while sinking was in progress, and then to
allow cement grout injection when the caisson
was at its final level. A composite concrete and
steel air deck was constructed above the tunnel
eyes. The central section was fabricated from
steel beams and plate for rapid installation and
removal.

20.   Compressed air was provided by mobile
containerized 56 m³/min compressors with containerized chillers. Power was also generated
by containerized generators and switchgear.
While containerization entailed more initial
capital investment, it allowed rapid installation
at any site. Experience showed that air losses
were minimal and generally at the lower levels

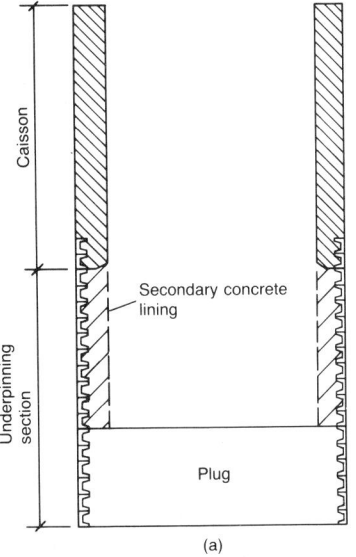
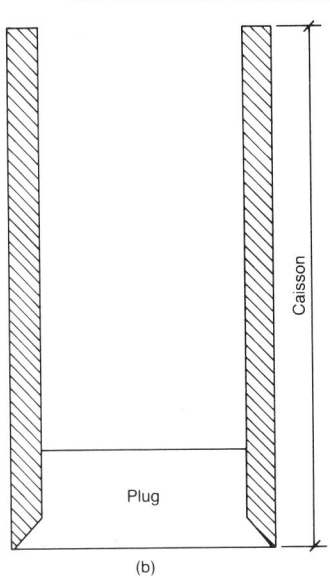

a sensitive balance was reached between the air
pressure in the working chamber and the down-
ward load.

21.   Excavation was carried out in the
working chambers by means of small
electrically-driven hydraulic excavators
loading into skips which were removed by
crane through the vertical air locks.

22.   Sinking rates were dictated by excavation rates from within the working chamber as
the caisson structures were built upwards in
advance of sinking (Table 2.)

23.   One or two shafts experienced excessive
resistance at times (probably due to collapse of
the annulus) and had on rare occasions to be
blown down. This was a strictly controlled
process carried out in the presence of engineering staff and entailed the sudden release of up
to 25% of air pressure for a period not exceeding 30 s from start of reducing of pressure to
restoring the hydrostatic balance. No noticeable
ground loss was experienced on these
occasions.

24.   The verticality of caissons was controlled in all cases by excavation and not by
distribution of kentledge.

25.   On contract 12 some of the caissons
were sunk in very narrow streets within touch-
ing distance of tall, old buildings of doubtful
construction (Fig. 4). These were all constructed without any distress to the buildings
and it is a commendation for the full-depth
caisson method.

Triple-cell machine launching caisson

26.   One exceptional feature of contract 3
was the construction of shaft 3 which was the
designated launch shaft for the two bentonite
tunnelling machines. This required major temporary works in the form of large diameter
underground chambers to enable the machines
to be launched in opposite directions in close
succession.

27.   To solve this time-consuming and
potentially hazardous operation the triple-cell
full-depth caisson idea was developed. The
caisson was elliptical, 30 m long and 13 m wide

*Fig. 3. Alternative
methods of shaft
construction;
(a) Engineer's design;
(b) Contractor's
design*

Fig. 4. Sinking of caisson in narrow street

*Table 2. Rates of
excavation: m³/h*

Average	4·1
Maximum (sustained rate during 8 h shift)	7·9

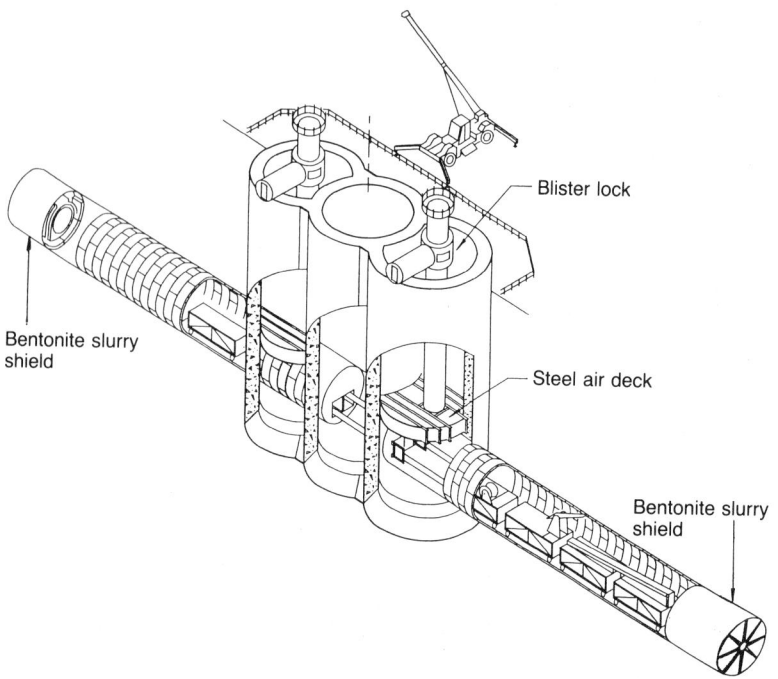

Fig. 5. Tunnelling from triple-cell caisson

Fig. 6. Triple-cell caisson

with a central cell of 10 m diameter. The arrangement enabled one machine to operate while the other was being assembled and commissioned in the same shaft bottom. The general concept is shown in Fig. 5.

28.   The complex shape of the triple-cell caisson required massive and intricate reinforcement to cope with the expected loading conditions, which at times created difficult steel fixing problems. The air decks for the outer cells were constructed in reinforced concrete with removable boiler lock sections bolted into central openings. The air deck of the central cell (the shaft proper) had to be sufficiently large to accommodate the insertion of the tunnel machine cutting head and bulkhead assemblies as complete units and was therefore removable. It consisted essentially of an immensely strong prefabricated steel platform.

29.   This complex structure was initially sunk in 'free air' to approximately 8·5 m below

ground level (Fig. 6). Thereafter sinking continued in compressed air without any undue problems down to the required depth of 24·5 m. Construction and sinking took 20 weeks, of which only the last five were in compressed air. A maximum of 2·07 bar was used together with a maximum air consumption of approximately 113 m³/min.

Slurry shields

30.   Two 180 t Northern Engineering Industries Limited machines (Fig. 7) built in the United Kingdom to Bade and Theelan designs, were used to drive the large diameter tunnels on contract 3. These machines had rotating cutting heads of the Berlin wheel type, which were essentially spoked wheels with adjustable vanes. In open mode these vanes allowed large boulders and obstructions through into the head chamber, and could also be closed down if required into a series of cutting slots, each 400 mm wide. The cutting heads were also provided with replaceable tungsten-carbide-tipped teeth. Face support and spoil removal was provided by pressurized bentonite pumped and returned through 250 mm dia. pipelines to a processing plant at the surface. Bentonite pressure was maintained within close tolerances using an upper compressed air chamber, to which emergency access was possible via a small air lock.

31.   A rubber tailseal system was employed to prevent ingress of water, ground and bentonite into the tunnel building area.

32.   Machine operation was not successful initially. Firstly bentonite pressures, spoil volumes and machine advance involved manually operating up to 130 controls for each ring-building cycle, and maintaining perfect equilibrium between all interrelated factors. This proved difficult in practice. It was not possible to achieve the preferred solution of a single integrated electronic control. The second problem was the ineffectiveness of the tailseal. Failure of the tailseal system required emergency replacement involving time-consuming and potentially hazardous operations.

33.   The tailseal system used initially comprised one operational seal mounted close to the trailing edge of the tailskin together with one emergency pneumatically inflated seal located inboard of it. The operational seal was triangular in profile and its apex was designed to trail along the back of the segments to create the seal. This failed to perform as required and resulted in a series of ground collapses ahead of the machine due to loss of bentonite pressure from massive leakage around the tailskin. Compressed air was applied to the whole tunnel to enable the machines to progress while a suitable replacement was found. This problem resulted in several months' delay.

34.   The final solution involved replacing the rubber seals with two rows of wirebrush seals with a special grease packed into and between them. The grease was replenished as necessary as the machines progressed. After this change progress improved dramatically and rates of 10–12 m per 12 h shift were achieved without further mishap.

Labels on Fig. 5:
Bentonite slurry shield
Blister lock
Steel air deck
Bentonite slurry shield

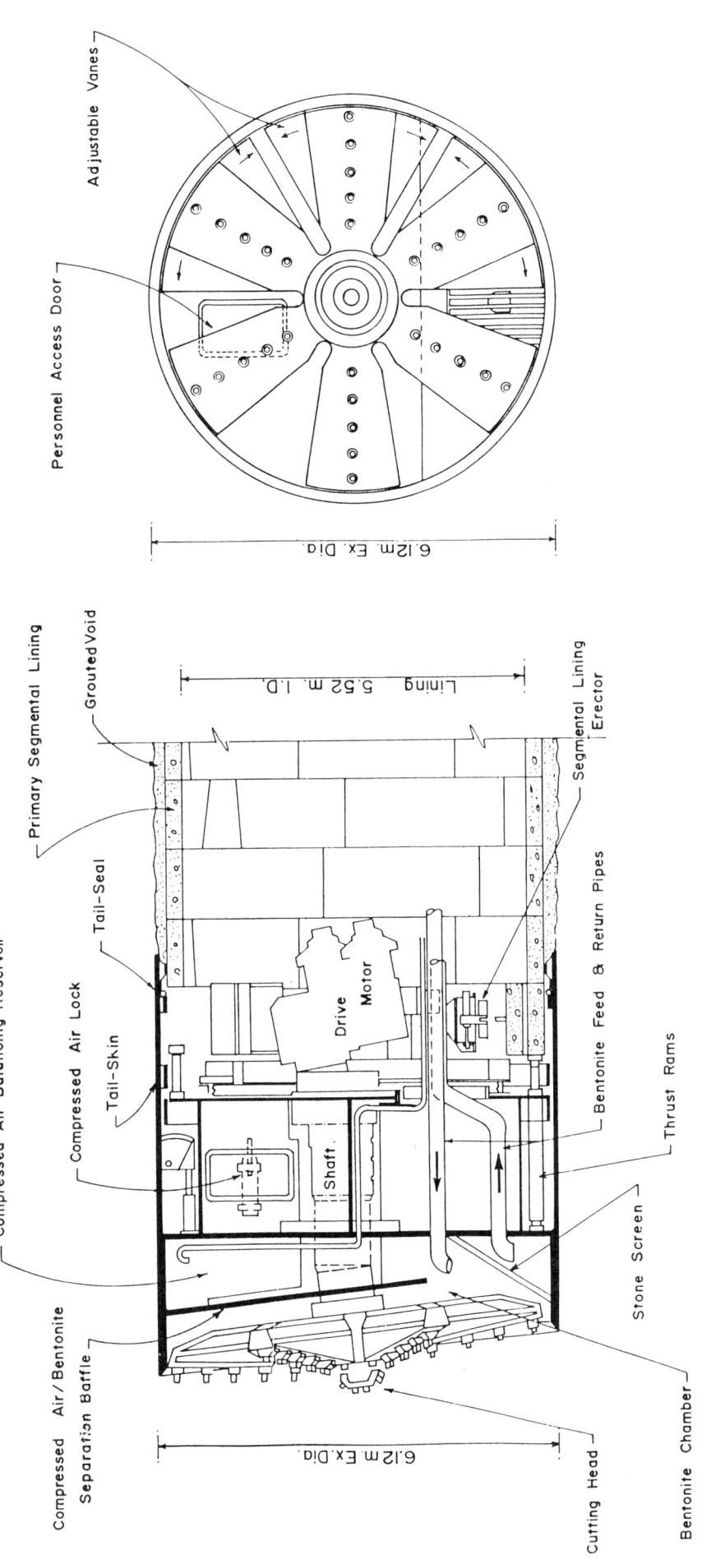

Adjustable Vanes

Personnel Access Door

6.12m Ex. Dia

Primary Segmental Lining

Grouted Void

5.52 m. I.D.

Lining

Segmental Lining Erector

Compressed Air Balancing Reservoir

Tail-Seal

Compressed Air Lock

Tail-Skin

Drive Motor

Bentonite Feed & Return Pipes

Compressed Air / Bentonite Separation Baffle

Shaft

Thrust Rams

Stone Screen

Cutting Head

Bentonite Chamber

6.12m Ex. Dia

Fig. 7. Main features of hydroshield used on contract 3

35. The segments of the slurry-shield-driven tunnels were fitted with EPDM rubber gaskets to make them watertight. While this worked satisfactorily, it required constant control of the ring erection to ensure adjacent gaskets were sufficiently in contact.

36. Grout injection was through screwed sockets fitted with plastic non-return butterfly valves. The grout consisted of cement and locally available limestone powder. The quantities injected were usually 50% more than the theoretical annulus.

37. As both slurry shields were fed from the same shaft it was decided to install bulkheads in each tunnel to allow compressed air to be applied to one drive (for example to change a tailseal) without disrupting the other. These bulkheads were formed of 10 N/mm² compressive strength sand/cement mortar with a thickness of 0·4 × tunnel diameter. Calculations showed the bulkheads would be in compression.

38. The completed tunnels were dry and brick lining was carried out using a series of gantries which moved along as a train. This allowed brick laying rates of around one million bricks per month per train. Epoxy mortar pointing was carried out by hand-held pneumatic guns.

Deformed section of tunnel

39. Approximately 1·4 km from shaft 3 on the downstream drive to Ameria, a 50 m length of newly constructed tunnel showed signs of deforming soon after leaving the machine tailskin. Examination revealed that the shape of the soffit was still approximately circular and had remained at the correct level, whereas the invert had heaved and distorted. This unusual development warranted a thorough investigation.

40. In places the heave approached 160 mm, with the lower quadrant of the affected rings taking up a scalloped configuration resulting in appreciable lipping between some segments. Despite this and the relatively high external groundwater pressures at the invert no leakage occurred, thus demonstrating the suitability of the specified gaskets under such adverse circumstances.

41. Immediate emergency remedial measures involved temporarily bracing the vertical axis of the distorted length with a system of massive whalings and struts until the cause could be ascertained and an appropriate permanent solution devised.

42. Investigations into the various factors involved revealed that on this particular length the grout had failed to harden, and by retaining fluid it had been unable to transfer the required lateral restraint as the rings left the tailskin. Also, it had subjected the lower quadrant of the rings to the maximum hydraulic upthrust at this critical time.

43. Rebuilding the distorted tunnel in situ was considered too risky and an internal strengthening solution was adopted involving heavily reinforcing the invert lining concrete and mechanically bonding this to the deformed rings by shear plates fixed to the bolts. This

was carried out in compressed air (to partially destress the tunnel) once continuous monitoring had confirmed that the situation had stabilized, and after the tunnel machine had completed its drive to Ameria.

Hand excavated tunnels on contract 3

44. The smaller compressed air tunnels on contract 3 were constructed in a conventional manner with hand excavation and Greathead type shields. High air losses led to the problems of cavity grout being blown away, large voids being formed behind the lining and a considerable amount of back grouting and caulking required on completion of tunnelling to achieve the specified watertightness.

45. In one tunnel drive where the sands at tunnel level were overlain by an extensive thick clay layer, an inversion phenomenon was experienced whereby an increase in air pressure led to increased water ingress in the tunnel invert. Research showed this not to be a unique experience and was probably caused by the entrapment of pressurized air leading to an accumulator effect.

46. Various methods of mitigating air loss were attempted, including pocket holing with clay around the crown of the tunnel face, spraying the face with grout, sealing with plastic sheeting and dewatering in the invert to allow lower air pressure. Flash grouting (injection of cheap, very rapid-setting silicate grouts through lances driven up to 1 m into the tunnel face) was possibly a method that should have been persevered with.

47. Average progress in these smaller tunnels was less than 30 m per week.

Main tunnel on contract 12

48. For the main tunnel on contract 12 it was felt that the slurry machine technology at that time would not cope with the ground conditions with a significant amount of clay and silt. Consequently, this was planned to be driven in compressed air using conventional shields fitted with a mechanical back-hoe. Generally these performed as expected, but air losses were much higher than anticipated. Losses of 560 m³/min on a 4·57 m diameter tunnel were experienced. In view of these losses it was decided to change the method of tunnelling in the northern section of the contract, which was in sandier ground, from compressed air to slurry shield. For this the contractor leased a Markham/Okumura bentonite slurry shield machine. The tunnel segments on this drive were fitted with hydrophilic gaskets, which were found to be more expensive, but were considered more effective than the corresponding EPDM gaskets on contract 3.

49. An interesting feature of contract 12 was the section of tunnel in rock at the southernmost end. Although comprising weathered limestone, the rock here proved too hard to be removed mechanically and blasting had to be used. This is one of the oldest and most crowded parts of Cairo. Blasting was carefully controlled within the vibration limits stipulated by the authorities. The transition of the rock head gave a long area of mixed face where com-

pressed air was used to stabilize the soil in the crown of the tunnel.

Settlement considerations

50. Settlement monitoring was carried out regularly on all tunnel drives, and the different face support techniques used on these two contracts produced some interesting results. Generally, settlement was within the predicted ranges, although on occasions it was in excess of, and on others it was surprisingly well below that expected.

51. Settlement on the contract 3 slurry shield drives was generally in the range of 30 mm to 60 mm, whereas on the two contract 12 drives (using the Markham/Okumura bentonite slurry shield) settlement was significantly less at approximately 30 mm on average, excluding bentonite control incidents such as those referred to earlier.

52. Settlement above the four large-diameter compressed air drives on contract 12 was 14 mm on average—i.e. substantially less than on the bentonite drives. The main reason for this is considered to be the immediate and continuous support provided by compressed air until the annulus was grouted up compared with the less positive support provided by the bentonite behind the shield, even though attempts were made to pressurize and stabilize the annulus by feeding it with bentonite off the main supply circuit. Settlement above the rock drive was very small at 3 mm on average.

Ground treatment

53. Two different types of ground treatment were used. On contract 3 in the sands of permeability around 10^{-5} m/s it was decided to use chemical (silicates) injection by the tube à manchette method. This was generally successful in reducing permeability to around 10^{-7} m/s. Although the effect of the treatment was noticeable, compressed air was still required in the treated zones. On contract 12 jet grouting was used, as chemical injection was not considered appropriate for the siltier material. Columns were drilled at around 1 m centres and grout injected at around 500 bars pressure. Some ground heave was experienced locally but rarely more than 200 mm. This treatment was more positive than the chemical treatment and, if anything, gave problems in excavation due to its hardness. Nevertheless there were some examples of 'windows' in the treatment and one of these gave rise to a blow out of the tunnel with the complete loss of pressure and a significant collapse at street level.

Conclusion

54. The full-depth caisson technique proved extremely successful without causing any adverse effects either on the surrounding ground or adjacent properties, and subsequently this same technique was used on the Boulac Spine Branch Tunnel.

55. Once the original tailseal problems with the Bade and Theelan designed hydroshields had been solved, they performed well and their increased production rates enabled the lost time to be recovered.

56. Conversely, the compressed air driven tunnels were generally more troublesome, where the open nature of the ground required much higher quantities of air than expected, especially on the larger tunnels towards the south of contract 12, and also overall production was only a fraction of that on the bentonite tunnels. The use of traditional caulking techniques with these tunnels proved less satisfactory than the gaskets used on the bentonite tunnels.

Main tunnel (contract 4)

57. Contract 4 on the project comprised the construction of 4·9 km of large diameter tunnels, 11 shafts, associated structures and connections to the existing wastewater systems between Souk El Samak and Abdeen (Fig. 1). The contract was awarded to the Lilley Construction/MISR Engineering Joint Venture in June 1984 in the sum of £35·6 million and £E57·5 million and construction commenced in December of that year.

Programme and design

58. The construction activities were analysed and categorized into six main events: diversions, shafts, tunnels, brickwork, finishings and ancillary works. In order to meet the tight programme, shaft sinking was scheduled to commence at week 12, tunnelling from week 53 to week 126 and with all ancillary works completed and the sites cleared by the end of the contract period of week 178.

59. The Contractor elected to sink the caisson shafts according to the Engineer's design and generally follow the recommended sequence in which the tunnel driving would be carried out by commencing at shaft 10. However, during the sinking of shaft 10, boulders were encountered within the horizon of the intended 4·0 m dia. tunnels to be driven upstream and downstream from this shaft, indicating that major modifications would be required to the cutter heads of both tunnelling machines. Due to the time taken to carry out the modifications, the programme of tunnel driving was re-arranged and the 6·15 m external diameter tunnel machine was the first to be launched, but this was not until week 80.

60. However, all tunnelling was completed by week 145 in a period of 65 weeks—two weeks less than the clause 14 programme.

Shaft sinking

61. Contract 4 comprised the sinking of 11 shafts as reinforced concrete cylinders, under compressed air and in very confined working areas. The shafts were sunk through fill deposits, silts into sands and the excavation completed in sand or gravel layers. The finished diameters of the structures varied from 6 m to 10 m and were designed as sinking caissons for the first 10 m depth and thereafter to be constructed using locally manufactured concrete segments by the underpinning method for a further 15 m. An in-situ reinforced concrete lining to the shaft segments was provided.

62. Figure 8 shows the specially designed caisson and lowering equipment with special

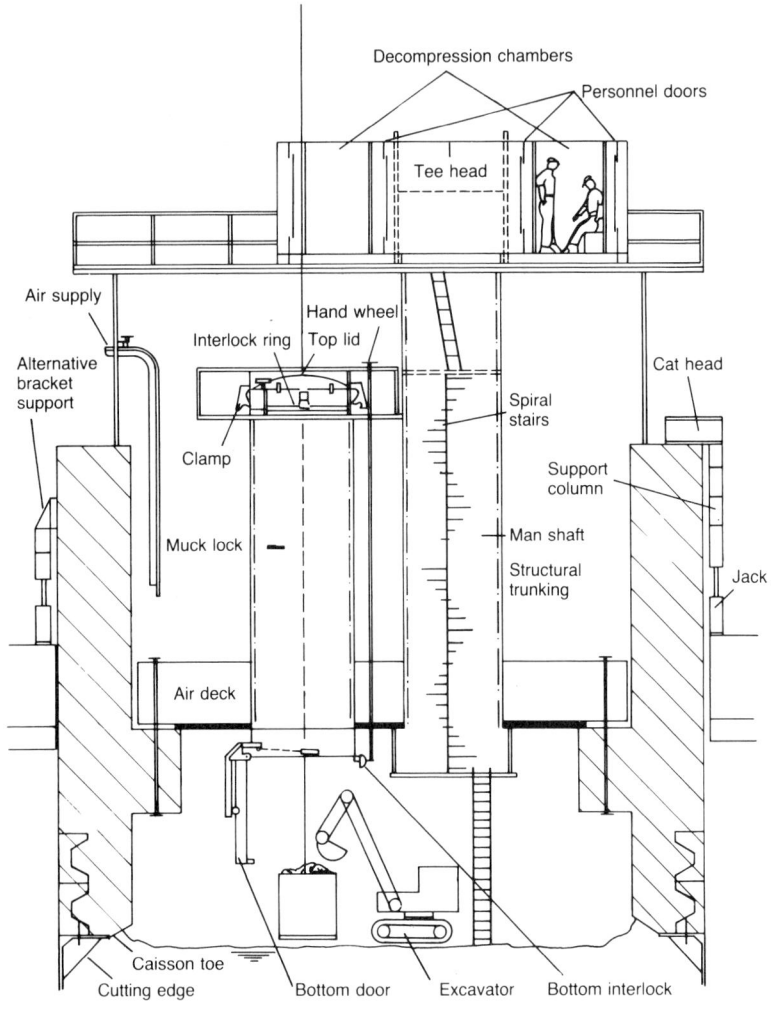

Fig. 8. Caisson air and lowering equipment

attention and emphasis paid to the excavation method, and access and egress for the labour to and from the working area. The design incorporates a decompression system which avoided decanting at the end of the working shift. Excavation of the shaft spoil was by a Smalley hydraulic excavator into 2·5 m³ skips and progress was generally of the order of 6 to 8 weeks for each shaft from application of the compressed air to pouring the 3 m deep concrete plug.

63. Due to some variation of the depths of the shafts, the low-pressure air varied from 1·6 to 2·3 bar requiring between 42 and 340 m³ free air delivery in the coarser soils.

Low pressure compressed air (LPCA)

64. All shafts and hand-driven tunnels were specified to be constructed under compressed air. Additionally, the slurry shields were required to be launched under compressed air, with a full back-up system of air available throughout the tunnelling period. This required the installation of bulkheads located approximately 100 m from each working shaft to allow a working area within the tunnel for pipework,

control systems, rail crossings, etc. Permanent compressed air installations of 250 m³/min were established at the two main tunnel-driving shafts (6 and 10) and a further two mobile compressed air units of 170 m³/min for sinking the caisson shafts. All electric power generation was by imported static and mobile generators totalling 12 000 kVA.

65. Two fully equipped medical centres were established on the contract to deal with all aspects of compressed air working and were continuously manned during the entire period of shaft sinking and tunnel driving, dealing with 41 700 decompressions over 1 bar resulting in a very low overall bends ratio of 0·74%.

66. The volumes of air used were small when sinking shafts through the silts and clays but increased steadily after sand was encountered, particularly during underpinning. One effect of the absorption of compressed air into the ground was to raise the groundwater level. Increases of up to 600 mm were recorded in piezometers adjacent to shafts during sinking. Several basements were flooded and one building rose and fell on removal of air by 30 mm. The slope across the building was very small and hence the structure was almost unaffected. At another shaft site, where there was some concern over possible damage to an adjacent mosque, an annulus of ground outside the shaft was treated before shaft construction commenced. As a result, air loss was reduced by 30% compared with the immediately adjacent shaft. These minor problems apart, shaft sinking progressed well.

Ground consolidation

67. The site investigation data indicated that the soils through which the tunnels for contract 4 would be driven would be predominantly sands and sands/gravels with increasing silt content over the southern half of their length. The water table was close to the surface, typically 2–3 m below ground level.

68. The contract called for chemical treatment of the ground to provide stable zones at all shafts for the exit and entrance of the shields, at all tunnels where excavation was carried out by hand methods, and in areas where sensitive structures were located. It was primarily intended that the treatment of the ground would be completed before the shafts were sunk. The subcontracted consolidation work involved 19 000 m³ of ground treatment and, 44 000 m of drilled holes, and was successfully carried out at an average rate of 400 m³/week. Ground treatment was undertaken using the tubes à manchette method at 1·2 m centres. The ground was initially injected with a cement bentonite grout to decrease the voids ratio before proceeding to inject with the sodium silicate based grout. Two hardeners were used: Dynagrout T and Rhone Poulenc 600. The former was not consistently satisfactory and was not used after initially treated zones were exposed.

69. The ground treatment had two unexpected adverse effects on tunnelling. Firstly, the UPVC pipes left in the ground, while suffi-

ciently brittle that they were expected to be broken with ease by the tunnelling machine, often suffered plastic failure in excavation and lengths of pipe blocked the slurry return pipe. This problem was reduced by installing a box containing a grillage which could be opened to remove the broken pipes. Secondly, the residual chemical in the ground reacted with the bentonite to produce foam which caused minor difficulties.

Bentonite slurry tunnelling machines

70. The mixed sands, silts and gravels under a head of up to 25 m of water could not be excavated without the use of a tunnelling shield combined with a means of controlling the groundwater. The use of an open-faced tunnelling machine in conjunction with low pressure compressed air would have resulted in considerable air losses into the coarser ground. Additionally, balancing the air pressure and the water table across the 6 m excavated face would have proved difficult. The decision was taken to purchase two 5·15 m and one 6·11 m external diameter slurry tunnelling machines from Markham, manufactured in United Kingdom under licence from Okumura at a cost of £6 million.

71. The machines featured a fully closed cutterhead capable of providing positive support to the face. Ground was admitted to the excavating chamber ahead of the sealed bulkhead through hydraulically operated slot gates designed to admit stones of up to 250 mm, although no such stones had been expected. When the machine was not advancing—for example at weekends, during maintenance or even during ring building—the slot gates were closed and the face supported. A system of paddles and guide bars inside the cutterhead channelled stones larger than 50 mm to a crusher. All of the excavated material was removed from the tunnel in suspension in the return bentonite line. At the surface slurry cleaning plant, shaker screens and cyclones removed all but the very finest fraction of the excavated material. The combined screening plant for the two machines working at the southern end of the contract incorporated a silt press to remove surplus fines. The press was only needed over short lengths of the tunnel and restricted progress when it was in use. The tunnelling machines incorporated copy cutters to enable the negotiation of curves, and compressed air locks in the bulkheads. In the event, these small air locks were not used. When face inspections were called for it was preferred to pressurize the whole tunnel rather than risk any loss of air from the small reservoir ahead of the bulkhead.

72. Following the discovery of boulders during shaft sinking, additional investigation holes, 1 m in diameter, were ordered at intervals along the line of the tunnels, which confirmed that the presence of boulders was confined to a 400 m length at shaft 10. Testing gave the strength of the flints as up to 620 MPa. The limestone was soft and would have presented no problems to the tunnelling machines.

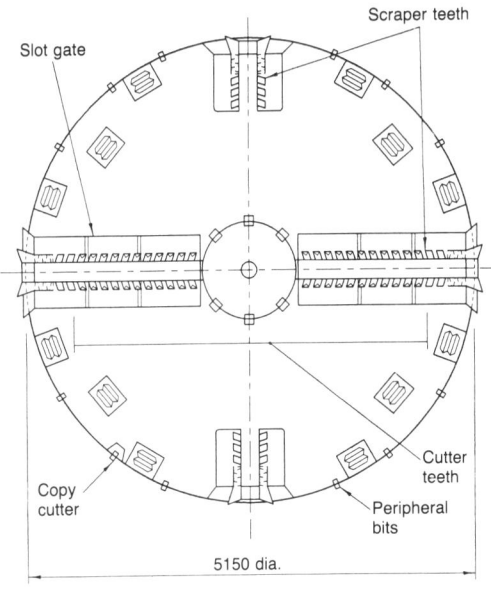

Fig. 9. Cutterhead

73. A decision was taken to modify the two 5 m tunnelling machines by providing disc roller cutters, replacing some scraper teeth with hardened cutter teeth and providing hard-faced welding on the cutter head (Fig. 9). In addition, peripheral cutter teeth and hard-faced welding were provided on the 6 m machine as a precautionary measure. The modified machines performed extremely well in excavating through the boulders.

74. The three machines were worked on double shift, six days per week, with maximum achievements of 108 m per week for the 5·515 m dia. machine and 168 m per week for the 6·11 m dia. machine.

Segments

75. The segmental linings, supplied by Charcon Tunnels from its specially-built factory located 40 km from the site, were of tapered design to eliminate the need for packings to negotiate the several curved sections on the contract and incorporated an expanding hydrophilic gasket to deal with the 25 m head of water. Both linings and gaskets performed well, producing a virtually watertight tunnel on completion.

Tailseals

76. The seals inside the tailskin of the tunnelling machines consisted of two rubber L-shaped seals and an outer wire brush seal. The rubber seal in particular suffered severe frictional damage moving across the outside of the tunnel lining, especially when hardened grout had built up against the outside of the seal. In order to protect the seal as much as possible, two measures were adopted: a wedge of soft foam rubber was placed immediately outside the seal to reduce grout build-up, and a softer grout consisting of cement, bentonite and limestone fines was used in preference to the originally specified neat cement grout. This proved to be relatively successful but on the two 5 m dia. machines working from shaft 10;

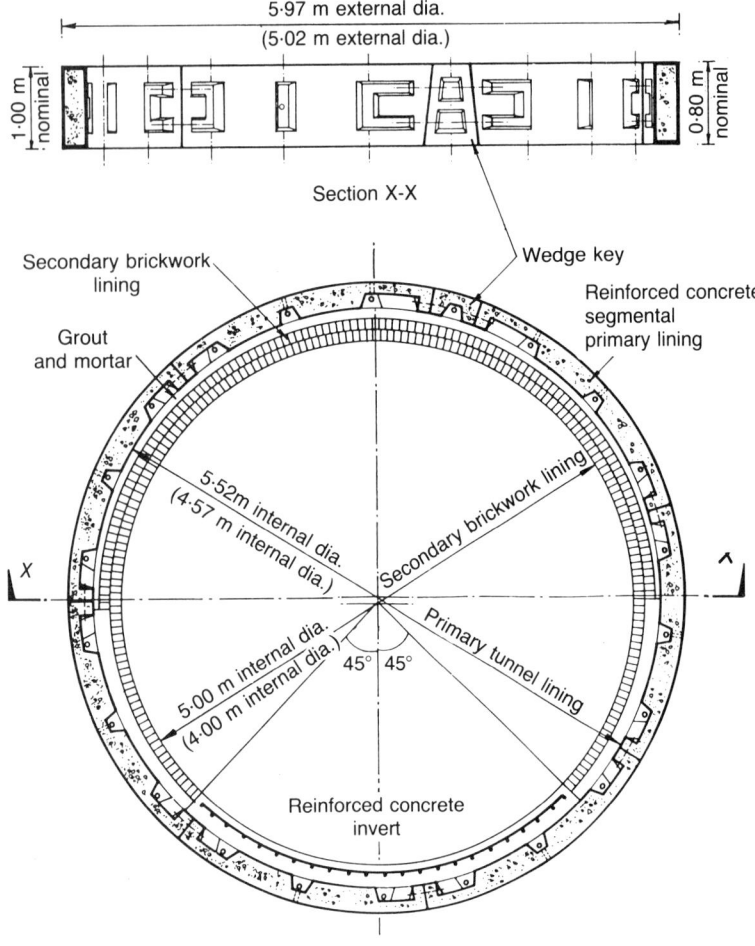

Section X-X

Fig. 10. Sectional elevation of tunnel and brickwork lining (looking forward)

the centre seal was changed from the rubber type to the more durable wire brush.

Settlement

77. For most of the length of the tunnels the settlement recorded did not exceed the predicted 20 mm for the 5 m dia. machines and 30 mm for the 6 m dia. machine, based on a 2% volume loss. Over isolated lengths settlement of up to 150 mm occurred and at some locations the road surface above the tunnel failed over an area of about 10 m². The recorded settlement exceeded that predicted when operational difficulties were encountered—principally bentonite pressure falling to as little as 0·5 m below groundwater pressure. In general, greater settlement was recorded where progress was slower and also at a length of tunnel designed with successive bends to avoid the piled foundations of an elevated roadway. The settlement monitoring confirmed that up to 25% of the settlement occurred ahead of the machine, commencing when the machine was approaching at a distance of some 20 m. The results indicate that the full extent of settlement had occurred once the machine had progressed 50 m beyond the monitored section, typically after three to four days.

Brickwork

78. The shafts and tunnels were designed for a minimum life of 100 years, and to guard against the aggressive nature of the effluent which generates hydrogen sulphide, all exposed surfaces were given a secondary lining of locally manufactured acid-resistant blue bricks and pointed with an epoxy resin mortar. All benchings and landings in the shafts received a corrosion-resistant screed and the soffits to the landings and shaft tops were cast with a PVC membrane.

79. In the tunnels the secondary lining consisted of a 90°, 200 mm thick reinforced concrete invert, a single skin of blue bricks to axis and a double brick arch from axis around to axis (see Fig. 10). The concrete invert was constructed using a moving shutter after the brickwork had been completed. In order to achieve this, concrete corbels were cast at each extremity of the bottom quadrant as the first operation in secondary lining. A cheek shutter was then set and backing concrete placed up to axis level. The acid-resistant bricks were laid up to axis level before ribs and lagging were fixed to support the arch of bricks during construction. The laggings were used to regulate the courses with a 600 mm gap left at the crown. A short timber shutter was then employed to support the bricks as the crown was closed. In order to ensure a mortar-free 25 mm deep pointing of acid-resistant epoxy mortar, the facing bricks received an adhesive wrapping 25 mm wide, and polystyrene was used to form the pointing groove.

80. The bricks to be laid on contract 4 totalled around 5 million and, while initial problems were encountered in achieving outputs and maintaining the high standards set for keeping the bricks clean to receive the epoxy resin mortar, these difficulties were soon overcome and outputs in excess of 150 000 bricks laid per week were being met.

81. On completion of the secondary lining to the tunnels, the bricking of the shafts followed in quick succession followed by shaft capping, installation of manhole frames and covers, restoration of diverted services as required, relaying of pavements and road surfaces and the return of the working sites to their former purposes.

Conclusions

82. The tunnelling work was particularly successful, confirming the wisdom of the choice of tunnelling method and machine. Despite the unexpected occurrence of large boulders the machines, once modified, showed that they could cope with a variety of ground conditions.

Boulac Spine Branch Tunnel (contract 12/5A)

Design

83. The new sewer relieves the grossly overloaded existing sewer system of the Boulac District, by intercepting it at ten existing pumping stations and connections, and makes provision in the form of blind adits for eight future branch sewer connections. The most

important connection is the existing overloaded
collector 3, which conveys the flows from the
Garden City and surrounding districts in the
centre of Cairo. The new sewer flows from collector 3 in Sharia El Galaa, westwards, then
northwards until it turns east and crosses
under Sharia Shoubra and Sharia El Teraah El
Boulakia and the Upper Egypt and Delta railways eventually discharging into the new trunk
sewer at shaft 4 constructed under contract 3
near Souk El Samak Pumping Station in the
north-east. Fig. 11 shows the route of the tunnel
relative to the main topographical features,
including the newly constructed trunk sewer
tunnel (contracts 3, 4 and 12), the railways, the
proposed Metro, and the Nile and its previous
courses in the past 1000 years.

84.   The connection requirements combined
with the need to route the sewer under streets
or open spaces to minimize settlement effects
on adjacent properties, required 20 shafts, of
which 10 are either vortex or combined
access/vortex shafts.

85.   The hydraulics were determined from
the levels of collector 3, the existing sewer connections and the need to be compatible with the
future Metro, originally intended to be constructed in open cut along Sharia Shoubra.
These criteria required invert depths ranging
from 10·0 m to 17·5 m, which, combined with
the sewer sizes, necessitated construction by
tunnelling.

86.   Pre-contract site investigations revealed
that in addition to the adverse ground conditions, the groundwater in Boulac is very high at
between 1·0 m and 2·0 m below surface level.

87.   Tender documents were based on compressed air being used for sinking the shafts,
driving the tunnels in conjunction with open-
faced hand shields or back-hoe mounted tunnelling machines, and constructing shield
launching and reception chambers, blind adits
for future branch connections, etc., in conjunction with ground treatment.

Contractor's alternative tender and method of construction

88.   The Contractor's tender incorporated
important and radical changes to the permanent
works and construction methods envisaged in
the tender documents. The Contractor proposed
that the size of all except one of the shafts
would be standardized at 6·0 m internal diameter and redesigned as homogeneous reinforced concrete structures to be sunk as full
depth caissons, as used successfully on contracts 3 and 12. The segmental tunnel lining
would be redesigned with ring lengths of 1·0 m
(nominal), comprising six almost identical trapezoidal segments with grooves for hydrophilic
rubber gaskets on the outer extremities of all
making faces. Rings would be tapered 15 mm
across their horizontal axes to enable negotiation of curves down to 200 m radius without
special joint packings. The essential features of
the redesigned ring are shown in Fig. 12.

Shaft construction

89.   All 21 shafts (20 permanent and 1 temporary tunnel-driving shaft) were constructed

as full-depth caissons. These incorporated two
important (albeit minor) changes compared
with those constructed on contracts 3 and 12 as
follows. All the tunnel services (i.e. air, water
and power supply pipes) were cast into the
caisson wall to bypass the air deck, thereby
enabling air decks to be installed and removed
quickly without such complications. Air decks
were circular and bolted to a concrete corbel
incorporated into the caisson wall, whereas on
contracts 3 and 12 the air deck was bolted to a
steel main frame which was pocketed into the
caisson walls. These modifications greatly simplified air deck operations.

*Fig. 11. Plan of
Boulac spine tunnels
and associated
features*

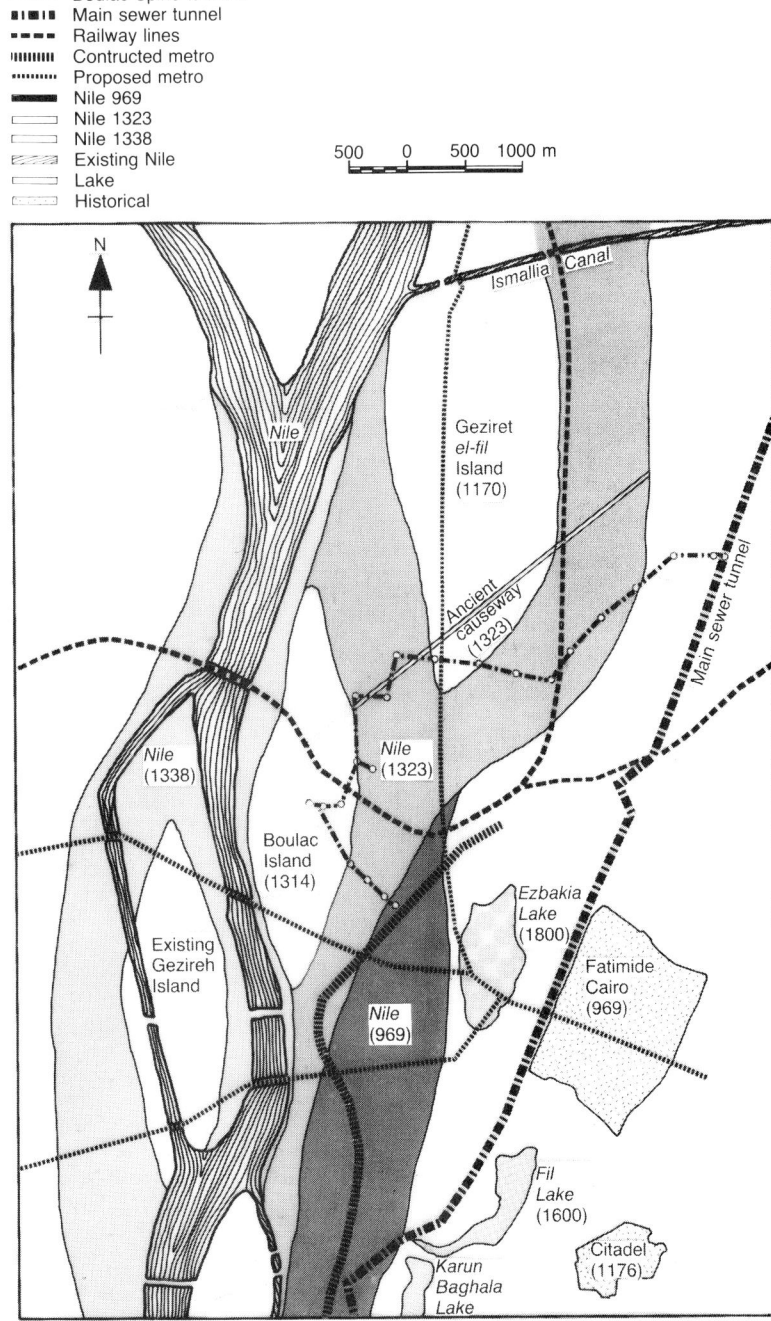

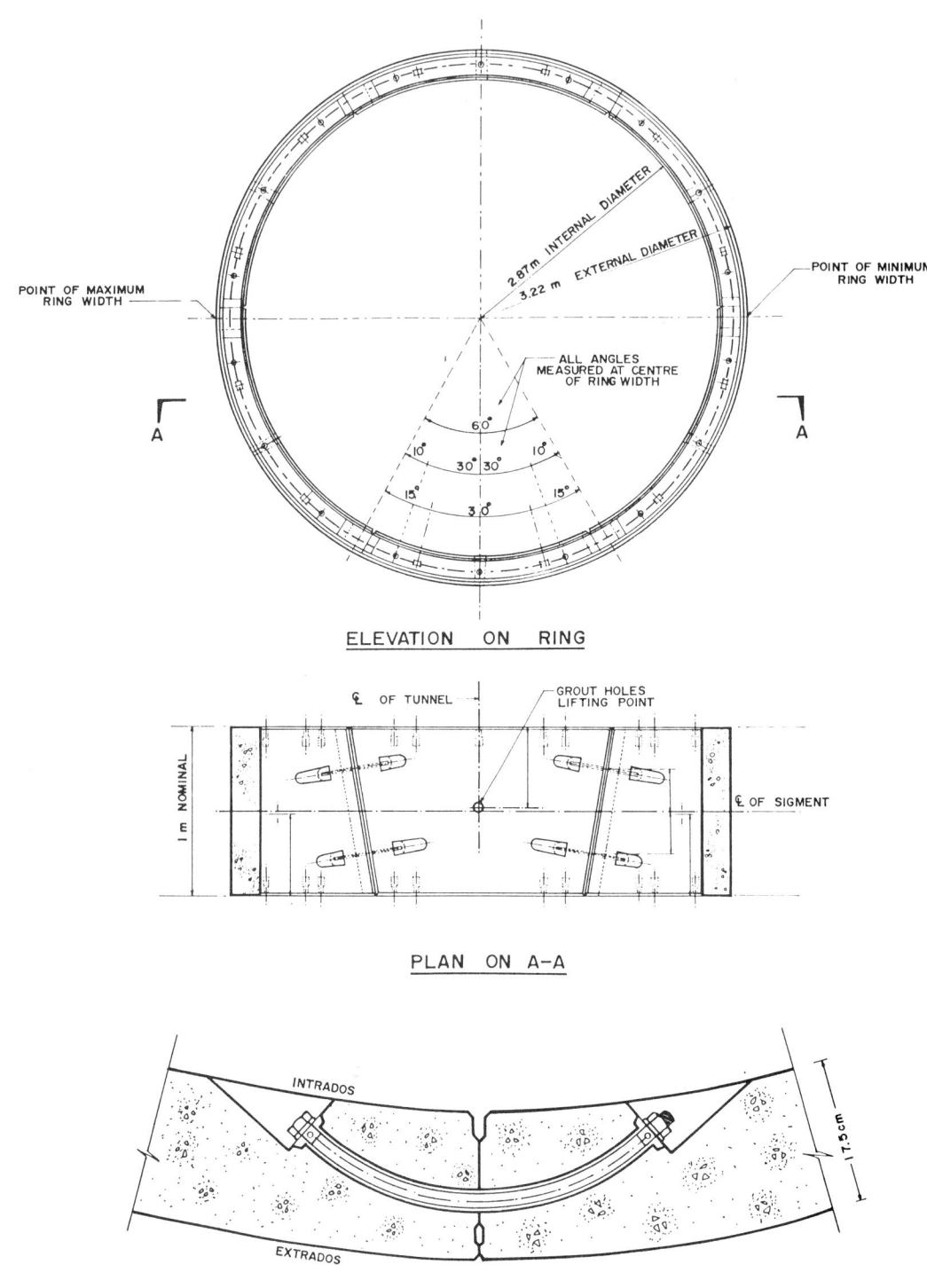

ELEVATION ON RING

PLAN ON A–A

SECTION THROUGH RADIAL JOINT SHOWING

CURVED BOLT CONFIGURATION

Fig. 12. General arrangement of revised trapezoidal segment

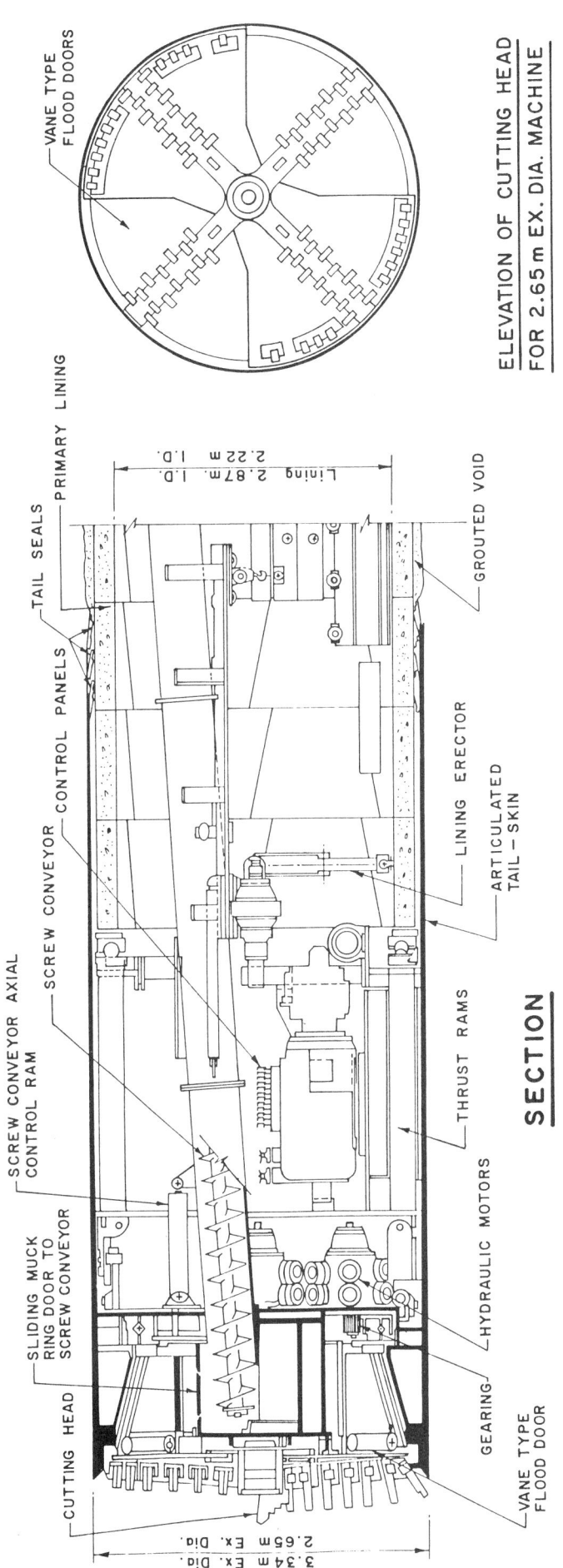

ELEVATION OF CUTTING HEAD
FOR 2.65 m EX. DIA. MACHINE

SECTION

Fig. 13. Basic configuration of earth pressure balance tunnelling machines

Earth pressure balance tunnel machines

90. Two Canadian manufactured Lovat earth pressure balance machines (EPBMs) of 3·22 m and 2·64 m external diameter were used for driving the tunnels instead of open shields and compressed air (see Fig. 13). The absence of bentonite type pipework, and hence of the continual need to extend it, simplified the tunnelling operation.

91. The EPBM shield housed a full-face rotary cutting head mounted on a bulkhead through which an enclosed screw conveyor extracted the excavated spoil in semi-stable condition for disposal in the normal manner. The very high excavating torques require peripheral motivation provided by four pinion motors driving a large-diameter gear ring running on the bulkhead bearing.

92. The tailskin incorporated triple wire-brush type tailseals similar to those on the bentonite tunnel machines. The bulkhead and tailseals combine to separate the external conditions from the tunnel, creating a 'free air' working environment. Segments are transported ahead of empty spoil skip trains, and rings are built with a hydraulic erector.

93. Successful operation required balancing spoil extraction, machine advance and thrust forces to maintain stabilizing pressures at the face, without over-excavation or over-stressing of segments. All such interrelated factors are controlled from a panel located just ahead of the ring building area and the operation required a high degree of expertise and skill.

Tunnel machine launch and reception

94. Within a shaft diameter of 6·0 m it would not be possible to launch the complete tunnel boring machine (TBM). To create more working space, chambers were constructed in LPCA at each break-out and break-in location. These were built from 'standard' Charcon bolted pre-cast concrete rings to an internal diameter just greater than the machine diameter, allowing for an all-round clearance of approximately 120 mm. These were cast in Egypt using imported moulds. The end of the chamber consisted of a two-ring (1·2 m long) bulkhead of 10 N/mm² mortar. The length of the chamber at 8·4 m was such that the head of the TBM would be buried approximately 0·6 m into the bulkhead when the first ring was clear of the tailskin, and it could be concreted into the shaft eye. An annular rubber seal was built into the launch chamber as a precaution against breaching of the bulkhead before the first ring could be sealed into the shaft eye. The void between the TBM and the chamber rings was grouted with a soft setting grout. The TBM then continued until sufficient tunnel had been driven to allow installation of the trailing equipment (approximately 70 m).

95. Thus the whole launch was carried out in free air. At the break-ins the problem was to avoid a run of ground occurring into the chamber as the TBM breached the bulkhead. This was done by pressurizing the reception shaft with LPCA. The length of the break-in chamber at 4·2 m was such that the TBM tailcan could be pulled off the last tunnel ring which was located within the bulkhead position and a concrete ring was constructed, sealing this last tunnel ring into the end of the chamber. Prior to the TBM pulling off this last ring the driven tunnel was pressurized from the drive shaft. On completion of the ring wall the LPCA was removed and the TBM split and turned and rejoined for the next drive.

96. Where upstream and downstream drives were in line no chambers were built and the shaft was partially filled with 5 N/mm² mortar (the eyes having been formed from 10 N/mm² mortar at the construction stage). After the TBM had driven through the shaft the mortar was excavated and the completed tunnel sealed into the shaft under LPCA with the tunnel intact across the shaft. Tunnelling continued uninterrupted. The completed rings were removed through the shaft at a later stage.

97. These methods worked very satisfactorily. The construction of a mortar-filled shaft adding, on average, two and a half 12 h shifts to the drive and break-ins took an average of 11 shifts, of which only half were in LPCA. Break-outs including 70 m of tunnel construction, averaged 20 shifts.

TBM performance

98. Commissioning and routine operation of the machines went as planned, and the ground conditions encountered were as expected. The only effects on progress were a very few machine breakdowns. After driving approximately 1 km, the larger machine's main bearing began to break up, and it had to be 'nursed' through to the next available shaft for removal of the machine and replacement of the bearing. Despite these isolated mechanical breakdowns both machines performed well.

99. The geometry of the tunnels dictated a different method of logistic support for each tunnel machine. With the larger machine (M131) drives the trains could pass each other on a crossing, enabling two trains to service the TBM. This was not possible on the smaller machine (M104) drives and therefore only one train could be used, which had to be emptied and restocked at the pit bottom and returned to the TBM. Cycle times were therefore longer. On all the M104 drive shafts except one, the alignment of the tunnels did not permit a backshunt. With this cycle, times were reduced by allowing

Table 3. TBM production rates: m/week

	M104	M131
Overall average from initial installation to final removal: m/week	64·8	70·3
Average production with complete back-up equipment: m/week	148·9	129·4
		143·2*
Peak production/10 h shift: m	20†	21
	16	
/12 shift week: m	181	209
/consecutive 12 shifts: m	192	222
/consecutive 4 weeks: m	675	679

* Excludes failed bearing delay effects.
† With backshunt.

a second train to be used. However, despite the different methods of working the performance of both TBMs was remarkably similar. In both cases the best production was achieved at the end of their respective sections when the crews had perfected their methods. The production rates for both machines were as shown in Table 3. The difference between the overall averages mainly reflects the greater number of changes in direction on the M104 drives.

Settlement considerations

100. The Boulac Tunnel is located in the recent alluvial deposits left as the Nile changed course to its present alignment (Fig. 11). Streets in Boulac are relatively narrow, and the majority of the older properties have been constructed by traditional means with few having the benefit of reinforced concrete framing. These factors, combined with the very high groundwater table, meant that settlement was a paramount consideration.

101. Settlement monitoring stations with cross-sections were established at regular intervals and verticality checks made on all tall and vulnerable buildings, minarets, etc., within the zone of tunnelling influence. Excavation quantities and grout injection per ring were carefully regulated to minimize settlement. Usually differential settlement effects were not sufficient to affect stability, although some of the older properties developed quite severe cracks.

102. Generally, settlement with the larger machine was 25 mm to 35 mm with troughs of 20 m to 25 m wide, whereas with the smaller machine settlement was normally 15 mm to 25 mm, with troughs usually 10 m to 15 m wide. Settlement was less predictable than on the trunk 5 m/4 m internal diameter tunnels, possibly due to a combination of shallower depths, younger ground and the different types of tunnelling machines used.

Conclusions

103. Within two years all 21 shafts were sunk by the full-depth caisson technique and 6·6 km of 2·87 m, and 2·20 m internal diameter tunnels were driven using the earth pressure balance machines through potentially very hazardous ground conditions highly charged with water under the narrow streets of the densely populated residential Boulac district of Cairo. Settlement was minor and generally insignificant. The tunnels are exceptionally watertight with no recorded leakage flows.

Reference

1. ELLIOTT I. H. and SPOLTON R. L. Greater Cairo wastewater project: design of east bank stage 1 works tunnels. *Proc. Instn Civ. Engrs*, Part 1, 1985, **78**, Aug., 807–829.

Proc. Instn
Civ. Engrs,
Civ. Engng,
Greater Cairo
Wastewater
Project,
1993, 34–47

Paper 10234

Conveyance system: pumping stations and culverts

J. F. Grimes, BSc(Hons), MICE, MIWEM, *A. E. Allsop*, BSc, MICE,
J. E. C. Henham-Barrow, FICE, FIWEM, *J. J. Long*, BSc(Eng), MICE,
and V. G. Thompson, FICE

*J. F. Grimes,
Divisional
Director, Binnie
& Partners;
Alternate Member
of Board of
Control, AMBRIC*

*A. E. Allsop,
Christiani &
Nielsen Ltd;
Project Manager,
Christiani-
Misr Concrete
J.V. (1984–92)*

*J. E. C. Henham-Barrow,
Acer Consultants Ltd;
Resident Engineer,
AMBRIC (1985–92)*

*J. J. Long,
Binnie & Partners;
Senior Resident
Engineer, AMBRIC
(1984–88)*

*V. G. Thompson,
Director,
Kier International
Middle East Division;
Project Manager,
Kier International Ltd.*

Introduction

In 1977 the conveyance system components were six major centrifugal pumping stations, approximately 80 km of pressure main and 23 km of open canal. This system conveyed a sewage flow of just over 1 million m³/day for partial treatment at Gabal el Asfar and Kossous. Disposal of the partially treated effluent to Lake Manzala on the Mediterranean coast was via the Kossous, Gabal el Asfar, Belbeis and Bahr el Bakr agricultural drains (see Fig. 1).

2. Prior to the commencement of the master plan study, the General Organization for Sewerage and Sanitary Drainage (GOSSD) had prepared studies and designs for new wastewater treatment plants at Berka and Shoubra el Kheima. Each treatment plant was designed to treat 0·6 million m³/day and the plants would serve the Nasr City and Shoubra el Kheima areas of Greater Cairo.

System components

3. The 1978 master plan study report[1] recommended that a new wastewater treatment plant should be provided at the sewage farm at Gabal el Asfar and that the Berka and Shoubra el Kheima wastewater treatment plants should be retained as integral components of the East Bank system. Two new pumping stations, a centrifugal lift station serving the proposed deep tunnel and a screw lift station serving the existing collectors, would be provided at Ameria. Some of the existing pumping capacity at Ameria would be retained for a number of years so that flows could be delivered to Berka wastewater treatment plant.

4. A major new conveyance system would be provided to deliver sewage flow from Ameria to the new treatment plant at Gabal el Asfar. This conveyance system would be a gravity flow conveyor which would allow direct gravity connections of sewer networks serving new populous towns along the route of the conveyor. To avoid excessively deep open cut excavation or expensive tunnelling, two screw lift stations would be provided at Kossous and Khalag on the 13·8 km long culvert conveyor which would end in a third screw lift station at Gabal el Asfar WWTP. Archimedean screw pumps would lift the flow at all these stations. A centrifugal pumping station would be provided at Kossous to transfer sewage flows to Shoubra el Kheima WWTP. The system would be completed by the provision of a branch culvert which would intercept flows in an existing collector north-east of Ameria. The first

stage system flow capacities are shown in Fig. 2.

5. During the design and construction stages, overflows have been incorporated into the conveyance system at two locations. These will act as discharge outlets to the Kossous Drain during the early years of operation until Gabal el Asfar WWTP is commissioned. The overflow structures are located at Kossous pumping station (PS) and at the interconnection chamber between Ameria PS and Kossous PS.

Future development

6. The first stage of the conveyance system is sized to convey peak flows up to 1·5 million m³/day from Kossous to Gabal el Asfar WWTP. At Kossous PS 0·35 million m³/day will be diverted from the culvert conveyor to Shoubra el Kheima WWTP. The first-stage system components have been designed to facilitate expansion of the system capacity to 4 million m³/day. Expansion will include the provision of

(a) a second tunnel PS at Ameria
(b) one additional screw pump in the collector PS at Ameria
(c) a twin 3·0 m × 3·5 m box culvert from Ameria PS to interconnection chamber (IC) 1
(d) a twin 3·0 m × 3·0 m box culvert from IC 1 to Kossous PS
(e) a second screw lift station at Kossous PS site
(f) a triple 3·0 m × 3·5 m box culvert from Kossous PS to Khalag PS
(g) a second screw lift station at Khalag PS site
(h) a triple 3·0 m × 3·5 m box culvert from Khalag PS to Gabal el Asfar.

Security of operation

7. A system risk analysis carried out during the design stage found that power failure at a pumping station would result in unacceptable sewage flooding. Failure at Ameria would be followed by flooding in the streets of Cairo. Failure at a screw lift station would result in sewage flooding because the existing agricultural drains do not have sufficient capacity to accommodate large overflows from the culvert conveyor. The analysis concluded that full standby power generator plant should be provided at each pumping station site to minimize the risk of flooding. The total installed generating capacity provided by diesel engine generators is 28·7 MW.

8. Failure alarms are installed at each pumping station and a dedicated telemetry system is provided which links the conveyance system pumping stations. Each pumping station monitoring centre is equipped with a transmitter/receiver station and these are interconnected by a telemetry cable run in a PVC duct alongside the conveyor culverts.

Pre-construction activities

9. The area of land acquired by CWO was approximately 60 ha, which was mostly made up of the 40 m wide site for the culverts. The greater part of the land was in private ownership and formal acquisition procedures were lengthy.

10. Approvals were also obtained from the Ministry of Irrigation for the drain and canal diversions and crossings, from the Cairo and Qalioubeya Traffic Departments for traffic diversions and access arrangements, and from several utility owners for the temporary or permanent diversions of utilities. An added complication in the site acquisition process resulted from the great number of residential buildings which had been hastily constructed between 1981 and 1984 on the culvert route in the areas adjacent to Kossous PS site. Arrangements for the acquisition of these buildings, which were demolished under the culvert contracts, included compensation and/or provision of alternative accommodation.

Contracts

11. Details of the various contracts awarded for the first phase of construction of the conveyance system are given in fig. 5 of reference 2. The works at Ameria were constructed under two contracts—one civil and one plant. By comparison, the Kossous and Khalag pumping stations were constructed under a single contract covering both civil and plant elements. All six conveyance system contracts were tendered by prequalified contractors. The pumping station contracts were restricted to UK firms or UK/Egyptian joint venture firms, and the culvert contracts were tendered by prequalified Egyptian contractors supported by UK firms who provided technical assistance. The locations of the contracts are shown in Fig. 1.

Ameria pumping station complex

Design layout of complex and its purpose

12. *Introduction.* The Ameria pumping station complex is the hub of the East Bank sewerage system. It receives wastewater from the central areas of the city through the new tunnel system and the old collectors. The sewage delivered by the tunnel is lifted by a new centrifugal pumping station, and that from the collectors by a new screw pumping station, into a gravity culvert which conveys the sewage out of the urban area for treatment and disposal.

13. *Principal flow routes.* Flows from the main tunnel collector pass through the distribution chamber (DC), via an interconnecting

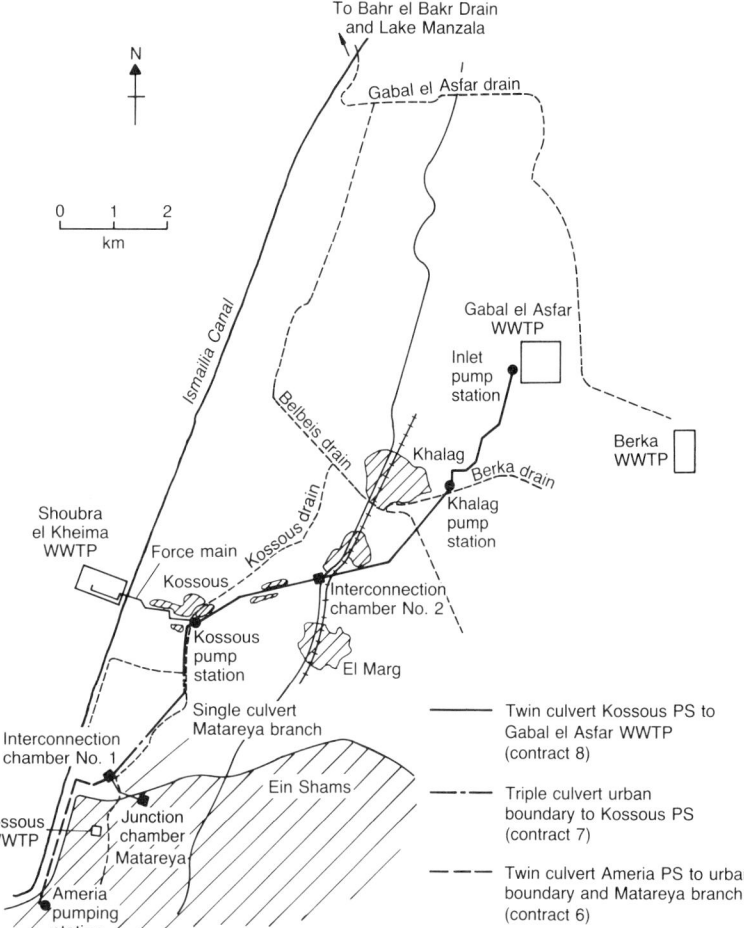

Fig. 1. Plan of conveyance system

tunnel, into the wet wells of the tunnel pumping station (TPS) (see Fig. 3). Eight centrifugal pumps lift the wastewater to the outlet channels which discharge to the headworks of the culvert system. Flows from collectors 2 and 3 and from the factories collector are diverted into the screw collector pumping station (CPS) which delivers via a culvert to the headworks of the culvert system.

14. There is provision for wastewater flows to be returned from the culvert headworks chamber to the wet wells of the four existing pumping stations for transfer to the new Berka WWTP in the short term, until sewers feeding that works from other parts of the city are completed. Bypass culverts and pipelines round both pumping stations are provided to convey wastewater in an emergency, if all power systems were to fail. In this case severe backing up of the collector sewers would occur.

15. *Major structures.* The TPS is designed for an average flow of 1·1 million and a maximum flow of 2·2 million m³/day. It has a 45 m external diameter and is 32 m deep (see Fig. 4). It has an internal dry well of diameter 28 m set 3 m off-centre to provide two annular wet well compartments. Flow through the interconnecting tunnel enters the wet wells forebay. Twin penstocks in each forebay/wet well dividing wall provide for wet well isolation, and the division wall separating the wet well compartments has a balancing penstock. The shape

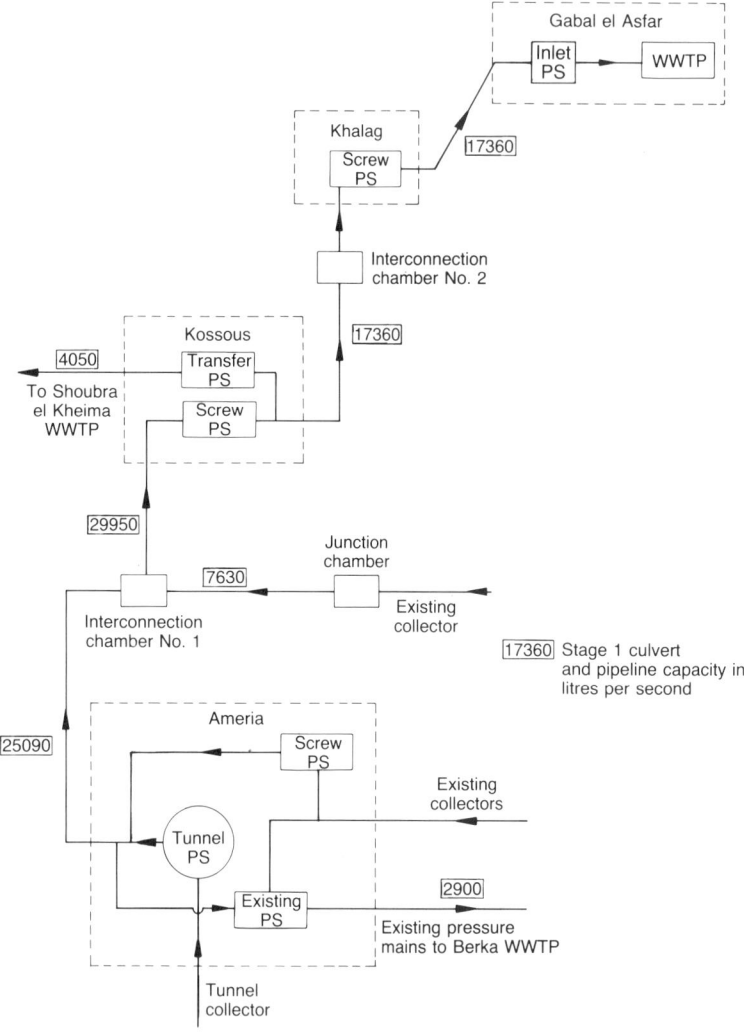

Fig. 2. Conveyance system flow diagram

the shaft sections are connected by rigid muff couplings, with the hydraulic and dynamic loads taken by thrust-bearing units fitted at motor and intermediate floor levels.

17. The DC is a circular reinforced concrete structure of 22 m external diameter and 32 m deep. It receives flow from the main collector tunnel which discharges over benching to drop 6·1 m to TPS wet well level. The flow through the chamber was model-tested to confirm that solids and grit would be transported through to the wet wells of the TPS, the TPS pumps being designed to pass a 150 mm sphere. Though screens are not provided in the DC, provision is made for coarse screens to be added if this is shown to be necessary as a result of operation. The DC will, in due course, split flow between the first stage and second stage TPSs. Provision for breaking in a 5 m dia. tunnel to the second-stage PS without affecting operation is made, and both the portals have 5 m dia. isolation penstocks.

18. The DC is linked to the TPS at low level by the interconnecting tunnel, which, though only 15 m long, is the deepest section of tunnel in the system.

19. The CPS houses four 2·96 m dia., 2·17 m^3/s Archimedean screw pumps (three duty, one standby) which lift the flow through 7·03 m. These are set at 38° inclination, operate at 25·46 rev/min, and are driven by 280 kW 3·15 kV motors through fluid coupling and a 56·53 : 1 gearbox. The design maximum flow is 0·75 million m^3/day. Provision is made in the civil works for a fifth pump, which is foreseen as part of second-stage development. The fifth channel was used to return flow during testing of pumps.

20. The CPS motor room, switchgear rooms and adjacent substation are founded on piles sleeved through the expansive upper clay layer to the sand/gravel strata beneath. The inlet culvert from the collectors and the inlet chamber are founded in the sand/gravel layers and are therefore not piled.

21. The main power supply is provided from a 66/10·5 kV transformer/substation. The continuous maximum demand load level for the Ameria complex is 15·3 MVA (stage 2 demand 20·4 MVA). Duplicate 10·5 kV underground cable feeders supply power to the TPS switchgear and control building from where power is distributed to all parts of the complex. Standby power is provided to permit operation of all systems in the event of power failure. The generating station houses four 3·73 MW diesel sets (three duty and one standby) which, on power failure, may immediately be switched in manually.

22. *Foundation conditions.* The foundation conditions for the TPS and DC are indicated by a borehole which was sited close to the centre of the TPS. This borehole had about 8 m of silty clay and made ground overlying dense to very dense sands and gravels to 9·5 m below TPS foundation level. Two thin hard grey silty clay layers at −10 and −13 m below datum

and suitability of the wet wells was tested in model tests at the British Hydromechanics Research Association for best flow conditions and grit transport, following which some shaping with benching was necessary. The model tests confirmed that the pump suctions were sufficiently submerged to avoid air entrainment. Grit removal facilities are provided for use if build-up is too great. Pump motors are located at ground floor level, connected by shafting to the pump units, which discharge to an annular collecting channel constructed above the wet wells. All concrete surfaces, which could be subject to possible attack by sewage gases, are protected with PVC sheeting keyed into the concrete face.

16. The principal features of the pumping plant in the TPS are

(a) 8 vertical end suction centrifugal pump units, each 3·6 m^3/s capacity, fixed speed 295 rev/min, pumping against a rated head of 22 m, and each weighing 35·5 t

(b) motors of rated output 975 kW, 10·5 kV wound motor slip ring induction pattern, cooled by ducted secondary air forced ventilation, and each weighing 40 t

(c) drive transmitted from motor to each pump by solid steel shafting enclosed in a steel support tube incorporating roller bearings;

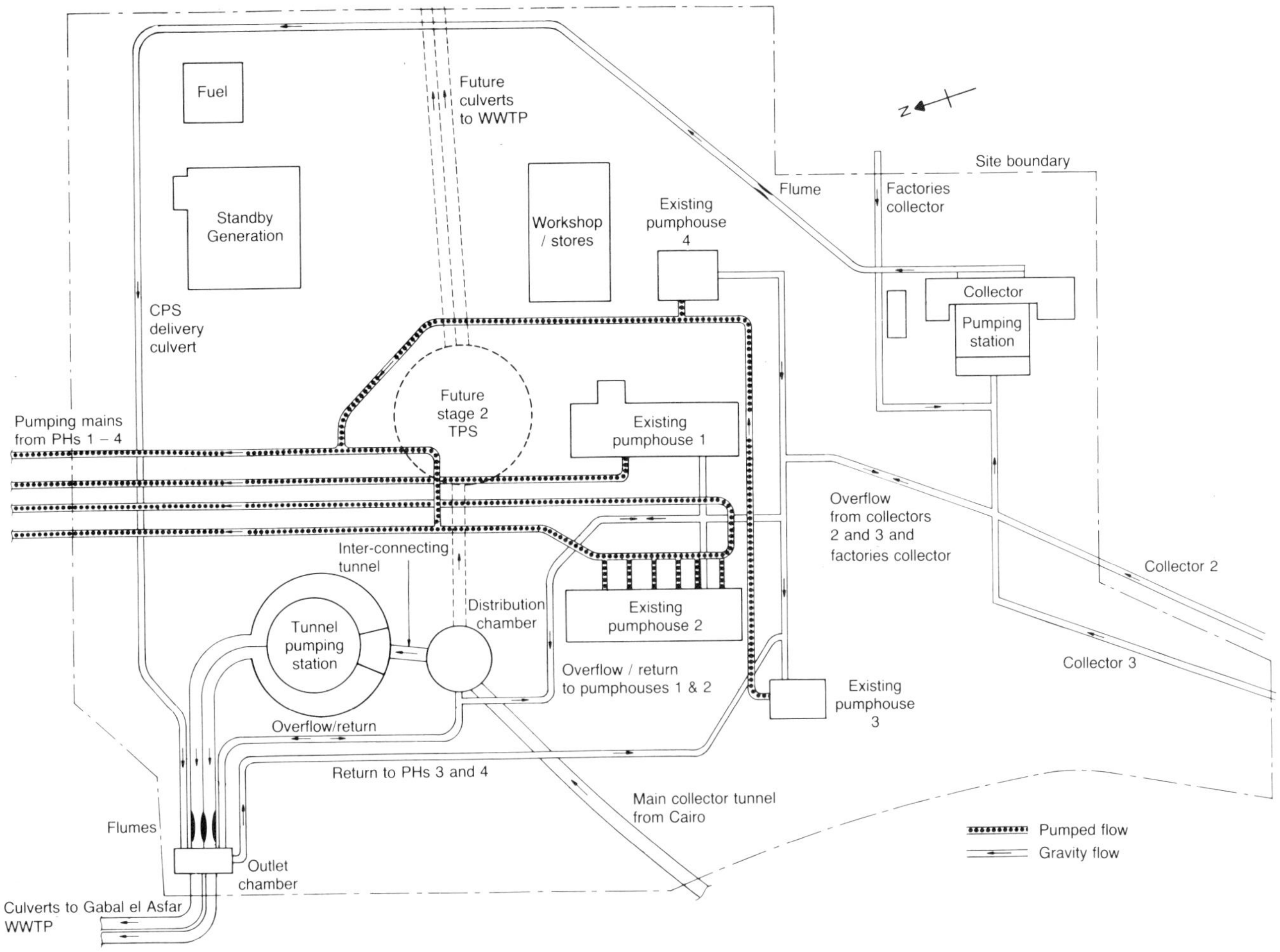

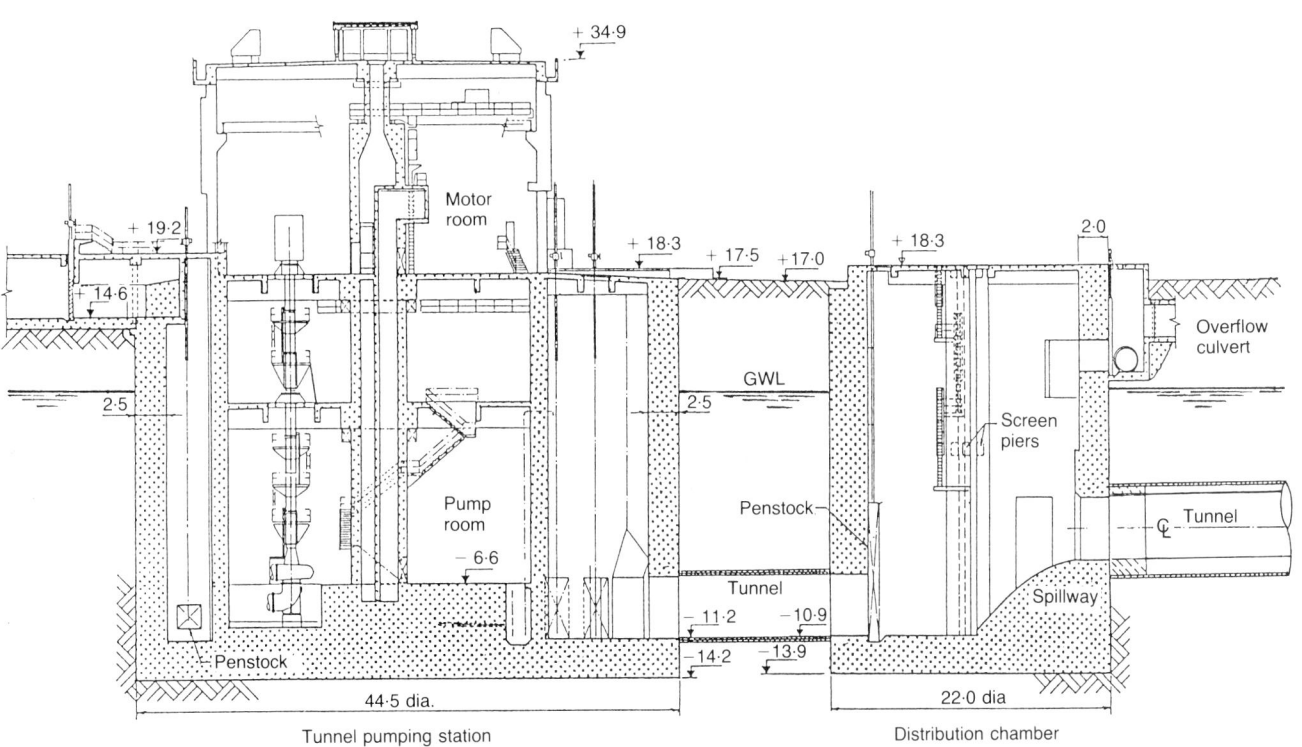

Fig. 3. General layout and flow routes

Fig. 4. Ameria tunnel pumping station and distribution chamber

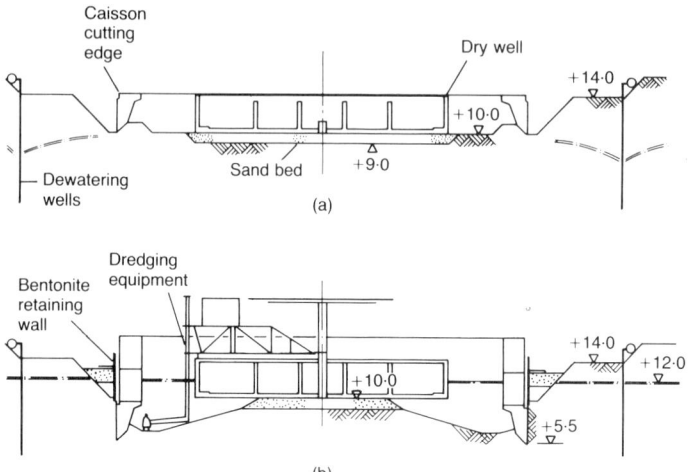

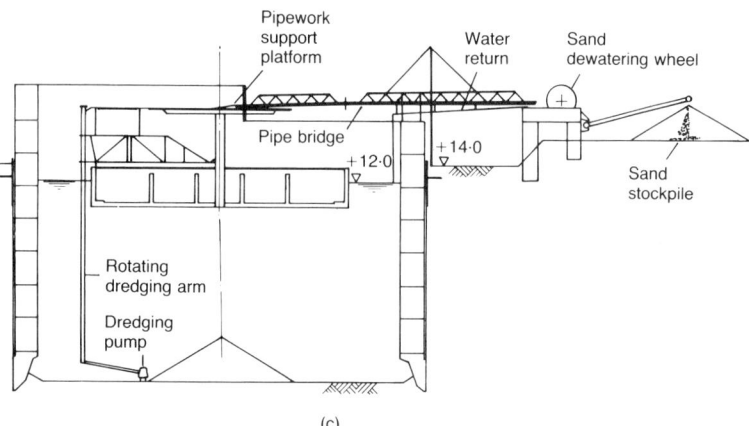

Fig. 5. TPS construction: (a) stage 1—construction of caisson wall with dry well base inside; (b) stage 2—wet excavation by dredging from floating dry well; (c) stage 3—wet excavation and caisson wall pours carried out simultaneously

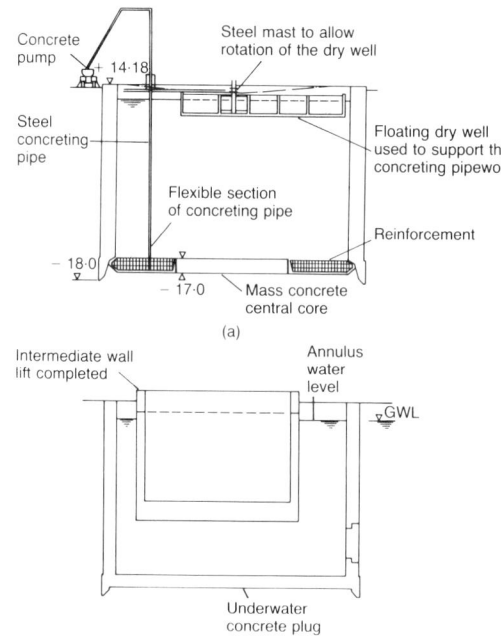

Fig. 6. TPS construction: (a) stage 4—underwater concreting of base plug; (b) stage 5—construction of dry well walls

were found. The sand/gravel strata are reportedly very deep. Foundation conditions on the remainder of the site are similar to those at the TPS. The upper layer of silty clay was shown to be expansive with an increase of moisture content.

Construction programme

23. The original programme for construction provided for the station complex to be completed and commissioned in 40 months. The work at Ameria was divided into two contracts: contract 1 for the civil work and contract 2 for the M&E work. Work started at the end of 1984 and the civil construction was complete in an extended period of 81 months, with M&E testing and commissioning taking a further four months. The times taken to sink the main caisson, to construct the interconnecting tunnel between the TPS and the DC, and the changed sequence and method of construction of the connection to the main collector tunnel, were the principal causes of the longer construction period.

Construction of tunnel pumping station

24. *Construction method selected.* No predetermined method of construction was imposed on tenderers. Possible construction methods were specified, such as ground freezing, compressed air construction, or the use of a diaphragm wall with a frozen base. The two lowest tenders both included proposals which involved the use of wet caisson sinking techniques. All methods of construction needed to avoid possible effects on abstraction at the water supply wellfield about 0·5 km to the west, and on low-cost multi-storey private buildings at the site boundary.

25. The method selected by the contractor, Christiani & Nielsen/Misr Concrete Development Co. JV, followed the stages illustrated in Figs 5–7. The outer wall of the TPS was sunk as a caisson (stages 1–3) using the base of the reinforced concrete dry well as a floating platform. An underwater plug was then cast (stage 4), concurrent with construction of the floating inner dry well (stage 5). The dry well was then landed on bearing points cast into the plug, and the void between base and plug was grouted. The dry well and wet well were then flooded with water ballast in turn, so that construction could be completed without risk of flotation (stages 6 and 7).

26. *Site clearance.* Space at Ameria was insufficient to locate the TPS without the diversion of major services and the demolition of some existing buildings. Two major pipelines carrying wastewater through the site from another pumping station crossed the TPS site. One was to have been diverted by others around the station before construction, but that diversion was not completed. The pipelines were redundant when the new tunnel collector had been commissioned, but temporary diversions were needed. Shutdowns were limited to 6 hours.

27. The older pipeline entered the site as two 36 in diameter CI pipelines, which joined to

leave the site as one 1·5 m dia. steel pipeline. The second pipeline was 1·5 m dia. steel. The diversions used welded steel pipework with saddle connections to the 1·5 m dia. steel pipework and a reinforced concrete boiler box connection to the 36 in CI pipework. Great care was taken to avoid potential gas hazards.

28. *Caisson sinking: the first stage.*   The cutting edge and cellular base for the dry well were cast at +10 m above datum. Reinforcement couplers were built in for continuity with the permanent base and the wet well dividing walls. The 150 mm annulus for bentonite formed by the cutting edge was subdivided into eight sections. Initially four alternate sections were filled with bentonite and four were sand-filled, to provide greater control during sinking. The bentonite mix was weighted with calcium carbonate to give a specific gravity of 1·3.

29. The outer TPS caisson was constructed in 18 equal pours of about 70 m³ each and 3·5 m high in a balanced sequence. The caisson started to sink in November 1985 with dry excavation using a small backhoe excavator working between the caisson and the dry well base.

30. During January 1986 the caisson was flooded, the dry well floated off, and the excavation started with sand dredging pumps suspended on dredging arms from the floating dry well base. Two DP50 Toyo pumps (each of capacity 300 m³/h) were used, each mounted individually on a hydraulically controlled (vertical and circular motion) mast and arm, mounted in turn on a hydraulically driven carriage to operate at any point round the circumference of the dry well base. It became evident that a larger pump would also be required, and in March 1986 a DP150 Toyo pump (of capacity 700 m³/h) was commissioned, and mounted in place of one DP50 but on a shorter dredging arm.

31. In the period to June 1986 caisson sinking continued, though generally in large single steps of up to 1 m with tilt up to 1° from vertical. During June bentonite was injected into the sand sections of the annulus below the sand. To promote sinking, excavation close to and at toe level was necessary. By July the caisson had sunk 15·6 m, with a further 12·4 m still to go.

32. In the period July to December 1986 the caisson was sunk a further 1·7 m, interrupted by a number of ground loss failures adjacent to the caisson bentonite annulus. In this period the alternate sand panels were cleaned out. The ground retaining the bentonite annulus between +4 m and +10 m was excavated and refilled with cement-stabilized fill, with the annulus face formed of steel sheeting with rebar tie-backs. Below +4 m the annulus was drilled out using contiguous 100 mm dia. drilling and bentonite return. The effective weight of the caisson was also increased by up to 25% with concrete block kentledge.

33. Following the ground failure in late December 1986, the method of sinking to both the TPS and the DC to final level was re-

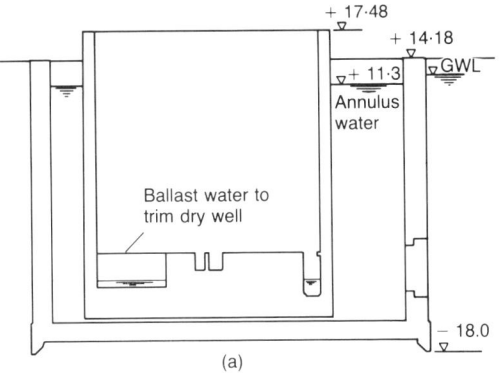

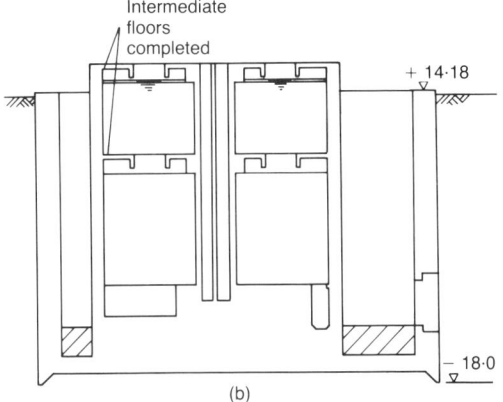

appraised. Sinking of the DC had continued in parallel with the TPS to a similar level.

34. *Caisson sinking: the second stage.*   The reappraisal resulted in the mobilization of rope-operated slurry trench grabs slung from gantries and winched hydraulically from gantries mounted to run circumferentially round the caisson wall top. The design of the gantries permitted excavation in a widened bentonite annulus and also inside the caisson close to the cutting edge, as illustrated in Fig. 8.

35. The bentonite used was imported from the UK and was mixed with water to form a gel controlled by a site test using a rod of 16 mm

Fig. 7. TPS construction: (a) stage 6—dry well landed on bearings and grouted under base; (b) stage 7—annulus is dewatered, ballast water added inside the dry well, permanent floor and radial walls added, then dry well dewatered

Fig. 8. Diagram of second-stage excavating equipment for TPS

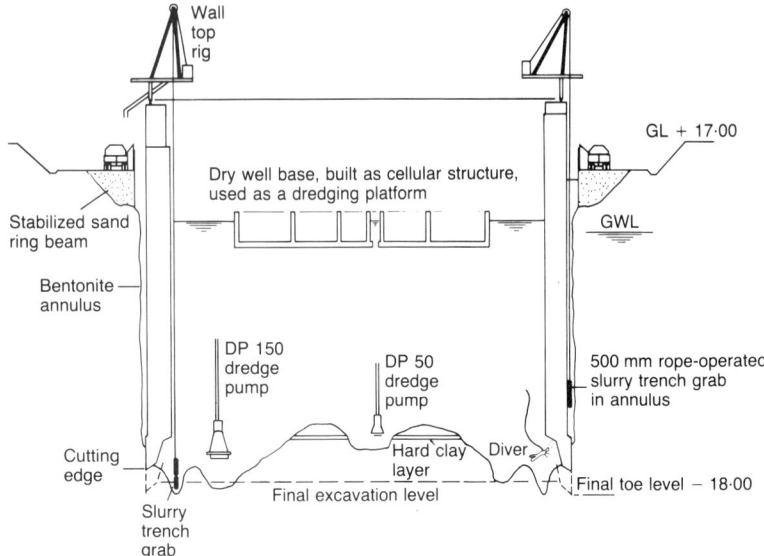

dia. penetrating 130 mm at most. The annular trench needed to stay open and stable for the period of sinking without change or recirculation of the bentonite. Remixing occurred owing to the action of the grabs.

36. Operations restarted with the smaller DC being sunk first. The 600 mm wide bentonite annulus was reformed and grab excavation began in early June 1987, following fabrication, shipping, assembly and commissioning of two grab units and gantries. The bentonite annulus was formed first to full depth, and then by the end of June sinking had been initiated. The DC caisson sank from −8·0 m to a final level of −18·0 m by the end of August 1987, the caisson finishing true to level, vertical and without measurable tilt.

37. Internal grab excavation, normally to 2·5 m below toe level, was needed to keep the caisson sinking. In addition a full-time diving team (4–5 men) was mobilized to excavate by water lancing along the inside face of the cutting edge to toe level. At one point the bentonite annulus was excavated 1·5 m below the top of the 3·5 m choker ring to promote sinking.

38. The two grab and gantry units were transferred to the TPS during September and a third unit was added. TPS sinking was restarted after excavation of the annulus in November 1987. The caisson was sunk to its final level by late May 1988, true to level and location, and within 0·1° of vertical. Safe diving operation during this final stage was paramount.

39. Sinking of the TPS was carried out similarly to the DC, with internal grabbing to 2·5 m below the toe and water lancing by divers, with a ground failure adjacent to the bentonite annulus occurring in November 1987. On two occasions during the sinking it was found necessary to excavate the bentonite annulus by up to 1·5 m below the top of the 3·5 m choker ring.

40. The full-time diving team also carried out detailed inspections to record internal profiles and faces, sometimes overhanging, which were often more than 5 m high, during the period when the cutting edge was between −10 m and −15 m below datum. A full-time diving engineer on AMBRIC's staff enabled surveys and sampling to be agreed as sinking continued. Concurrent with sinking by grabbing, bulk excavation was dredged and pumped to a dewatering wheel in which solids were sieved out and stockpiled and the water was returned to the caisson.

41. *Underwater plugging.* As soon as the DC caisson was sunk to its final level it was locked in position, tremie concrete being used to replace bentonite in the lower part of the annulus and sand elsewhere. The underwater temporary works plug was cast as a single 3·0 m thick underwater pour of about 800 m³. Hydrocrete was used as a concrete additive to improve the self-levelling characteristics and to avoid segregation. The pour was successful, except at one point on the contact with the caisson wall where a leak, evident during initial dewatering, was sealed by the diving team.

Subsequently the permanent floor was cast in the dry, followed by the remaining permanent works.

42. The TPS underwater plug, cast after locking the caisson in a similar manner to the DC, was technically more challenging. The central unreinforced circular pour of the plug under the dry well was cast first using precast concrete L-shaped forms which were left in position. The remaining annulus was divided into quadrants by four of the segmental reinforcement cages as dividing walls. Each of these dividing walls had a reinforcement cage to which was welded a framework of steel angle. Polypropylene cloth, with a woven texture designed to contain the concrete but leave an adequate construction joint surface, was stretched over the framework. The four dividing walls were then concreted. Individually tailored reinforcement cages (eight per quadrant) were then set in position and the four quadrants were concreted in turn. The divers set three steel plates to act as landing points for the dry well base.

43. *Dry well construction, landing and grouting.* The floating construction of the dry well ran concurrent with plug construction. The cells of the base were concreted according to a strict pattern as the walls rose. Complete circumferential pours for the walls for each 1·2 m lift were cast using a purpose-made climbing formwork system.

44. Landing of the dry well base was achieved precisely in location, level and orientation, bedding on purpose-made solid lead bearings. The void under the dry well was successfully grouted through pipes radiating from the centre of the dry well, and pumping continued until there was evidence of grout flow at all points around the dry well base.

45. The TPS structure was dewatered strictly in planned sequence without recourse to remedial action, and the remaining elements of the substructure were constructed.

46. *Superstructure construction.* Superstructure construction followed sequentially with TPS delivery channel construction. The radial reinforced concrete roof beams and parts of the roof were precast to advantage. As soon as the building was weathertight and the dry well was dry plant erection by the M&E contractor, CEGELEC, started.

47. *Plant installation and testing.* Erection of the TPS pumping plant and associated pipework was carried out using the permanent radial crane, which has a 25 t main hoist capacity and a 3 t auxiliary hoist. The TPS loading bay is sized to take the largest unit of plant on a low loader. During installation the presence of the sewage pumping main, which was still in operation at ground level between the TPS and DC, meant that CEGELEC needed to provide a temporary railroad with winched bogey between the main and the TPS motor room.

48. Plant installation was carried out in parallel with the civil works in the wet wells, since it was necessary to maintain overall progress.

49. Testing, prior to commissioning, had

logistical challenges, particularly in the routeing of the large volumes of test water required. As work was continuing in the DC, a temporary bulkhead was installed in the interconnecting tunnel, plus the leaving of temporary openings in the discharge channel floor to permit the testing of several pumps operating together. The wet well water levels were kept as low as possible to avoid overstressing the temporary interconnecting tunnel bulkhead.

Interconnecting tunnel construction

50. Ground freezing was adopted as the most suitable method for constructing the 15 m long interconnecting tunnel between the DC and the TPS (see Fig. 9).

51. In April 1989, ahead of TPS wet well dewatering, horizontal freeze tubes were installed in a pattern concentric with the interconnecting tunnel from the DC. Holes drilled through 'stuffing boxes', were cored through the DC wall. Through these, steel outer freeze tubes fitted with bits were drilled into position until contact was made with the TPS caisson wall. Chilled brine at −25°C, provided by an 80 000 kcal/h refrigeration unit (with a similar unit on standby), was circulated through the closed system to promote ice growth between individual freeze tubes and to form an integral horizontal ice cylinder. In mid September 1989, following eight weeks of primary freeze, free water pressure readings in the core of the tunnel ice cylinder, monitored through the reinforced concrete bulkheads built into the TPS and DC walls, confirmed ice closure. Following opening of the DC bulkhead, hand excavation of the tunnel progressed ring by ring.

52. By late October 1989 the heading had advanced 7 m, with six rings built. Each complete ring was excavated, erected and backgrouted in a two-day four-shift cycle. Preparation work to demolish the precast beam TPS tunnel bulkhead had started when ground was lost above the crown of the ice cylinder, which led to erosion of the ice cylinder, loss of the tunnel face and flooding of the tunnel, TPS and DC. Fracture of the 1·5 m dia. wastewater pumping main, diverted temporarily between the TPS and the DC, exacerbated the situation.

53. The loss of ground was accompanied by the fracture of 2 (out of 32) freeze tubes in the crown and loss of brine into the ground. The freeze system allowed isolation of the fractured tubes and restoration of freeze in the remaining tubes. Restoration included probes to determine the extent of the lost brine and the installation of a vertical freeze system to provide an ice 'hood' over the ice cylinder, with a further 150 000 kcal/h of refrigeration capacity.

54. Emptying of the flooded DC started in May 1990, and tunnelling in solid frozen ground was completed in early October 1990. The delay to plant installation in the TPS caused by the ground loss was 96 days.

Connection to the main collector tunnel

55. The connection design had provided for the construction of a tunnelling machine reception chamber, 7 m in diameter and filled with

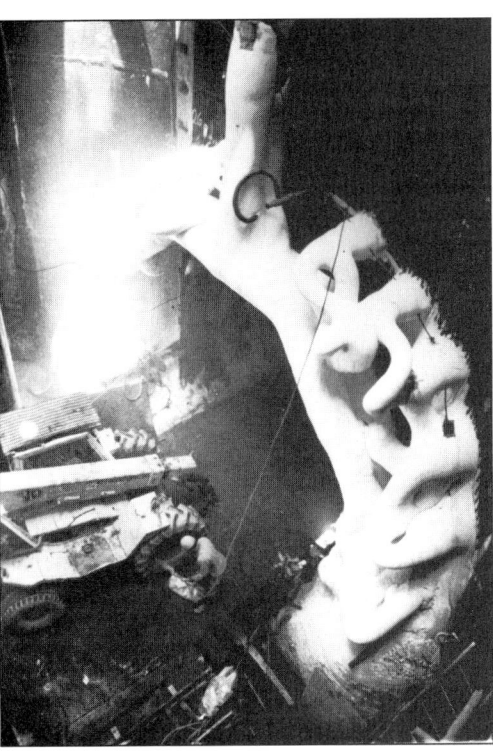

Fig. 9. Ground freeze at DC portal of interconnecting tunnel

weak grout, to be constructed from the DC. In the event the contract 3 tunnelling machine reached the DC before the chamber could be built. Connection to the main collector tunnel had therefore to be made with the tunnel boring machine (TBM) shield, tailskin and the mechanism forward of the TBM bulkhead remaining in the ground some 3 m from the DC wall.

56. Horizontal freeze techniques, supplemented by a vertical freeze system, similar to those on the tunnel were installed. The outer skin of the TBM was to be left in the ground with the other parts cut away.

57. The connection was completed by hand, including the removal of the TBM cutting face, gearbox and bulkhead, and the transition from circular to horseshoe lining was concreted 13 weeks after the establishment of the primary freeze, in March 1991.

Collector pumping station and connecting culverts

58. The base of the 9 m deep excavation for the CPS forebay was in saturated sands/gravels. A major system of deep wells was installed around the forebay and along the feed culvert from the collectors. The forebay and screw channels were constructed concurrently with the motor and switchgear/control rooms, which were founded at a higher level on piles.

59. The connection of the feed culvert to the three existing collectors presented special problems. Wastewater flows (2–3 m³/s in each) could neither be stopped nor bypassed outside the site nor could the surcharged levels be reduced. Detailed step-by-step planning and rigorous attention to safety aspects were necessary to achieve the connections under these conditions. Temporary works which accommodated the shape variations and allowed connec-

Fig. 10. Kossous pumping station site layout

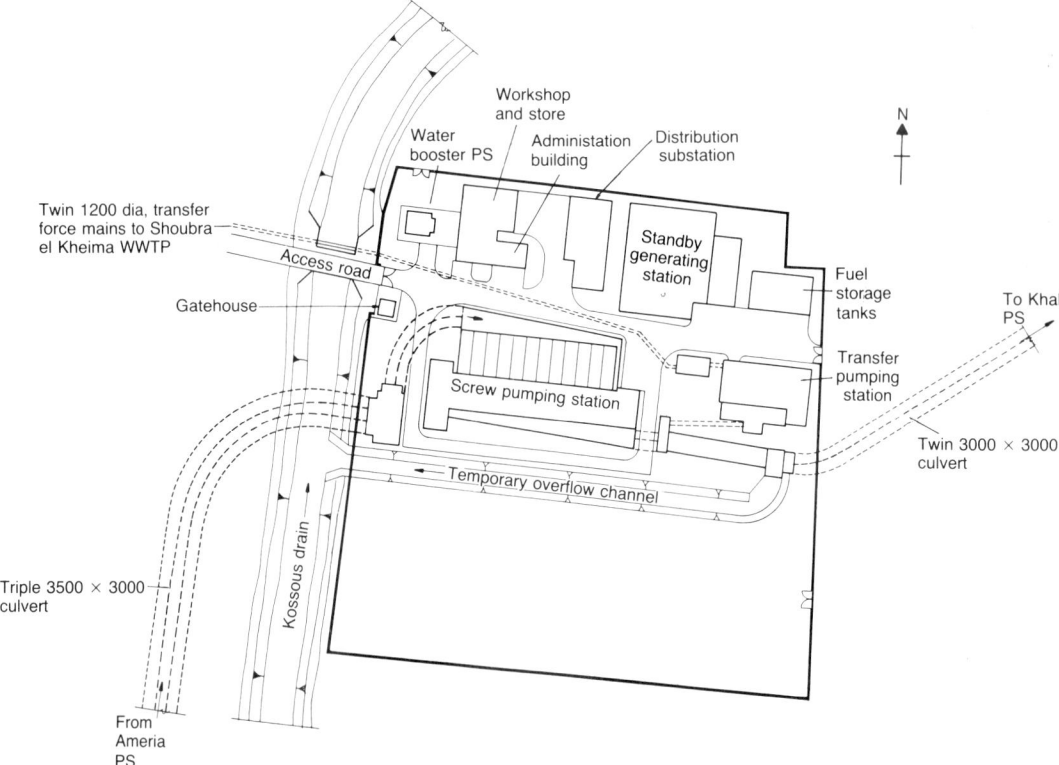

tions to be made in adequately dry conditions were designed and built.

Commissioning

60. *Initial operation.* Flows from sewer collector 2 were diverted into the CPS for three days starting on 13 January 1992 to check the downstream system. Diversion operations were confined to daytime (10 hours), with flows reverting to pumphouses 1–4 at night.

61. A programme of connections into the new tunnel collector, described in detail in reference 3, has ensured that wastewater flows were received at Ameria TPS in quantities suited to the pump sizes. On 16 January 1992, following connection at shaft 4, flow of about $3.8 \text{ m}^3/\text{s}$ was routed to the tunnel pumping station, enough for one pump, followed step-by-step with increased flows. The station was commissioned and on-line by 21 January 1992.

62. During this trial period the factories collector continued to feed to existing pumphouses 1–4. On 9 February 1992 the factories collector was diverted into the CPS. The diversion of these collectors into the CPS has increased their capacity substantially, as the new pump inlet is lower than those of the existing pumphouses.

63. By the end of July 1992 the TPS was discharging, on average, $14 \text{ m}^3/\text{s}$.

Kossous and Khalag pumping stations

64. Two lift pumping stations were required on the conveyance system in order to avoid excessively deep construction. The pumping stations are 6 km and 11·5 km downstream from Ameria PS. The first pumping station is located at Kossous adjacent to the

Kossous drain and the second is at Khalag, adjacent to the Berka drain. At each pumping station the flow is lifted 7·2 m. At the Kossous site a transfer pumping station has been provided to pump flow through pressure mains to the Shoubra el Kheima WWTP.

Principal features

65. The pumping station and support service installations at each site include

(*a*) a screw lift pumping station
(*b*) an electricity distribution substation
(*c*) a standby power generation station
(*d*) an administration and workshop building
(*e*) a reception chamber
(*f*) an outfall chamber
(*g*) a water supply booster pumping station.

Additional facilities at Kossous PS site include

(*a*) a transfer pumping station
(*b*) a transfer chamber
(*c*) a flow measurement chamber
(*d*) an overflow to Kossous drain.

Other features include access roads to both sites and pressure mains from Kossous to Shoubra el Kheima WWTP.

66. The largest structures are the pumping stations. The screw pumping station at Kossous is 79 m long, 45 m wide and 11 m deep to foundation level at the inlet bay, and the screw pumping station at Khalag is 70 m long, 45 m wide and 11 m deep. The Kossous transfer PS is 32 m long, 25 m wide and 9 m deep.

Design capacities and principles of operation

67. The Kossous lift pumping station contains twelve Archimedean screw pumps

(10 duty and 2 standby) which lift flows of 1·85 million m³/day from the incoming culvert from Ameria PS. Wastewater can be diverted from the screw pumping station outlet channel via the transfer chamber to the transfer pumping station. Five centrifugal pumps (four duty and one standby) in the transfer PS pump 0·35 million m³/day of flow through twin 1200 mm dia. and 2·5 km long ductile iron pressure mains to Shoubra el Kheima WWTP. The capacity of the temporary overflow to the Kossous drain is 0·65 million m³/day. A layout of the Kossous site is shown in Fig. 10.

68. The Khalag lift pumping station contains ten Archimedean screw pumps (8 duty and 2 standby) which will lift flows of 1·5 million m³/day from the incoming culvert from Kossous PS.

69. Three diesel engine generators provide an alternative source of power at each site: 8100 kW at Kossous and 7500 kW at Khalag.

70. The screw pumps at both Kossous and Khalag are started and stopped by manual control according to wastewater levels in the station forebays using a local control push-button adjacent to each pump motor. Transfer PS pump operation may be either automatic or manual. Normal operation is automatic, according to the wastewater level in the station inlet sump.

Construction

71. The pumping stations at Kossous and Khalag were constructed under a single contract which included the civil works and also the mechanical and electrical works. Kier International was the main contractor, with Nasr General Contracting Company (Hassan Allam) employed as the subcontractor for civil works and GEC Projects Ltd as the subcontractor for the mechanical and electrical works.

72. *Excavation, piling and groundwater control.* At the Kossous site a 1·5 m layer of made ground overlies approximately 4 m of silty material, which in turn overlies dense sand with traces of gravel. The soil conditions at Khalag site consist of a layer of dense sand to a depth of approximately 7 m, overlying an 8 m thick layer of silty clay, which in turn overlies dense sand with traces of gravel. At both sites the groundwater table is approximately 2 m below the surface.

73. When planning the dewatering system it was recognized that a perched water table existed on both sites, which if drained could produce undesirable artesian pressure and result in the danger of heave. The problem was overcome by the installation of sand drains around the perimeter of each excavation.

74. Deep wells drilled to a level below the clay layer were therefore able to deal with groundwater both above and below the clay layer. Around each of the pumphouse excavations 20 deep wells and 40 sand drains were installed. The deep wells, each with a capacity of 100 m³/h, reached a depth of 26 m, and the 600 mm dia. sand drains were 17 m deep. The deep wells were connected to a common header pipe which discharged into nearby drains. All

pumps were energized from generators, and round-the-clock maintenance ensured the continuous running of the system, which was required for a total period of 18 months, and allowed the excavation work to proceed satisfactorily.

75. Piling was required at both sites for the screw pumping station motor and control room, for the transfer PS control and switchgear room at Kossous and for the standby generation station at Khalag. Piling was carried out prior to the commencement of excavation. Precast concrete end-bearing piles were used at both sites. The 350 mm × 350 mm concrete piles were cast on site and installed using a piling rig with fixed leaders and a steam hammer which developed 3·5 tm per blow. Some problems were experienced initially in penetrating the clay layer with this method, and jetting techniques were used to penetrate the clay to within 1–2 m of the design depth in the underlying sandy material. The piles were then driven to a final set. Test piles were loaded to 120 t to confirm the adequacy of the piling system. In total 860 piles were driven and the pile length varied from 17 to 23 m, with piles founded 15 and 9 m below the foundation levels at Khalag and Kossous, respectively.

76. In general the sides to excavations were battered, but in areas where space was restricted sheet piles were driven a depth of 2 m below foundation level and anchored using ground anchors.

77. *Concrete works.* Concrete for the structures on both sites was produced by a batching plant located at the Khalag site, which consisted of two Elba EMC 35 plants, each capable of producing approximately 30 m³/h and fed by four 100 t silos and adjacent aggregate bins. Chilling facilities were installed and concrete was transported by insulated 6 m³ truck mixers. Most of the concrete was placed by Schwing concrete pump.

78. Shuttering was made up of standard panels either in steel or timber. A particular feature of the shuttering work was the requirement for a PVC lining to all surfaces exposed to wastewater.

79. *Pressure mains.* The transfer pressure main to Shoubra el Kheima WWTP comprised 2·5 km long twin 1200 mm dia. ductile iron pipes. The pipelines passed under the Ismailia Canal through a 3·2 mm dia. tunnel which had been provided under a separate GOSD contract. Installation of the pipes in the tunnel was made more difficult because the tunnel had also to accommodate twin 500 mm dia. pressure lines which would be used for pumping sludge from the Shoubra el Kheima WWTP. All four pipelines were installed on precast concrete plinths which were fixed at 4 m centres. In all 46 plinths were required, and these and the pipes were manhandled into position using basic hand winches and skids. Elsewhere the pipes were protected by plastic sleeving and laid on a prepared granular bed.

80. *Mechanical and electrical installations.* Details of the major items of plant installed at the two sites are given in Table 1.

Table 1. Major items of plant

Pumps		
Type	Archimedean screw	Vertical spindle end suction centrifugal
Number	12	5
Capacity: l/s	2170	1350
Diameter: mm	2960	—
Lift/head: m	7·2	15·4
Speed: rev/min	25	485
Manufacturer	Simon Hartley	Sulzer (UK) Pumps Ltd
Pump motors		
Output	280 kW	
Voltage	3·15 kV, 3 ph, 50 Hz	
Type	Cage rotor induction	
Manufacturer	GEC Large Machines Ltd	
Generator sets		
Equipment	Engines	Generators
Number	3	3
Speed: rev/min	1000	
Voltage	—	10·5 kV, 3 ph, 50 Hz
Output ratings	2·5 and 2·7 MW	2·5 and 2·7 MW
Type	Diesel 12 and 16 cylinder Vee form	Salient pole brushless
Manufacturer	Ruston Diesels Ltd	GEC Large Machines Ltd

Table 2. Culvert conveyor dimensions

Section	Length: m	No. of boxes	Size of each box (width × height): m	Overall size (width × height): m
Ameria PS to IC no. 1	2920	2	4·0 × 3·0	10·12 × 4·88
IC no. 1 to Kossous PS	3140	3	3·5 × 3·0	13·28 × 4·52
Kossous PS to IC no. 2	2360	2	3·0 × 3·0	8·12 × 4·45
IC no. 2 to Khalag PS	3140	2	3·0 × 3·0	8·12 × 4·45
Khalag PS to Gabal el Asfar PS	2290	2	3·0 × 3·0	8·12 × 4·45
Matareya branch	890	1	2·75 × 3·0	4·21 × 4·22

81. Installation of the screw pumps, each 18 m long and weighing 24 t, posed a problem because they had to be lowered into the prepared troughs at exactly 38°. Two mobile cranes were used to install the screw pumps. A 250 t capacity crane, with unequal length slings, placed the screw pump at the required angle (see Fig. 11). A 25 t crane was used to assist in lifting and moving the screw until it was over the trough. The screw pumps are covered with GRP covers.

82. Kossous screw pumping station received the first wastewater flow from Ameria on 13 January 1992. The controlled flow of 0.65×10^6 m³/day was discharged to Kossous drain; this arrangement will continue until Gabal el Asfar WWTP is ready and Khalag PS can be commissioned.

Culverts
83. The conveyance system from Ameria PS to Gabal el Asfar WWTP is a gravity flow conveyor, which includes two pumping stations which lift flows, thereby avoiding excessive and uneconomic depths of construction. At the end of the conveyor a third station lifts the flow into the treatment plant at Gabal el Asfar. The provision of a gravity flow system enables sewers from local communities along the route to be connected directly to the conveyor.

Main components
84. The 13·8 km long culvert conveyor from Ameria PS to Gabal Asfar WWTP comprises sections of twin and triple box culverts and includes two interconnection chambers, IC no. 1 and IC no. 2. IC no. 1 is located between Ameria PS and Kossous PS and IC no. 2 is located between Kossous PS and Khalag PS. At each pumping station the upstream culvert ends in a

Fig. 11. Screw pump installation

reception chamber and the downstream culvert starts from the outfall chamber. A 890 m long single box culvert, the Matareya branch, joins the culvert conveyor at IC no. 1. The main dimensions of the culverts are presented in Table 2.

85. The interconnection chambers, and the reception and outfall chambers at each pumping station are provided with penstocks which will enable individual boxes of the culvert to be isolated in 3 km long sections for inspection and maintenance. The interconnection chambers are sized to accommodate the future stage of the culvert conveyor; IC no. 1 is 31·3 m × 21·4 m × 12·5 m deep, and IC no. 2 is 27 m × 17 m × 8·8 m deep. In total 18 access points are provided. The covered access chambers are located approximately 1 km apart and are large enough to allow small mechanical loaders to be placed inside the culverts. Details of a triple box culvert and an access chamber are shown in Fig. 12.

Design features

86. The culvert conveyor is designed to carry at least 200 mg/l of sediment at minimum design flow, and the design allows for a sediment depth of 300 mm and 100 mm freeboard at peak design flows. This has resulted in gradients of 1 in 1900 for the twin box culvert sections and 1 in 2050 for the triple box culvert sections. The single box culvert has a gradient of 1 in 1625. A longitudinal section of the culvert conveyor is shown in Fig. 13.

87. The box culverts are cast in situ, reinforced concrete, water-retaining structures designed to BS 5337. All sections of culvert are designed to withstand hydrostatic uplift with the water table at ground level and also to withstand internal surcharge conditions. Construction joints are provided at 15 m intervals in an expansion–contraction, contraction–expansion joint sequence. Corrosion protection

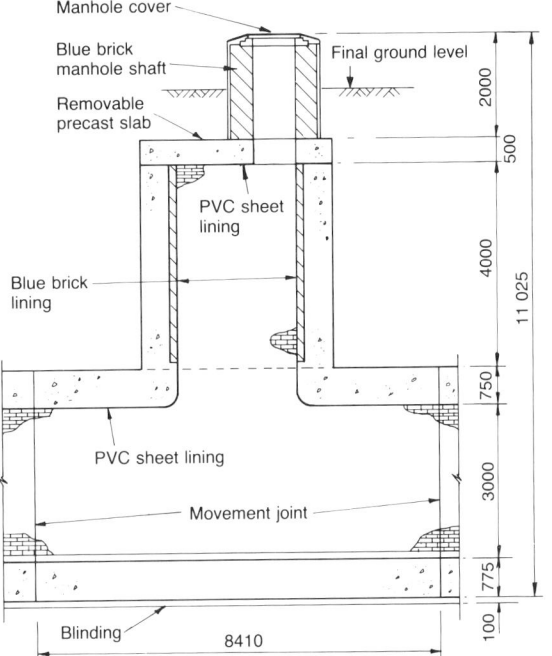

Fig. 12. Details of triple box culvert and access chamber

Culvert access chamber

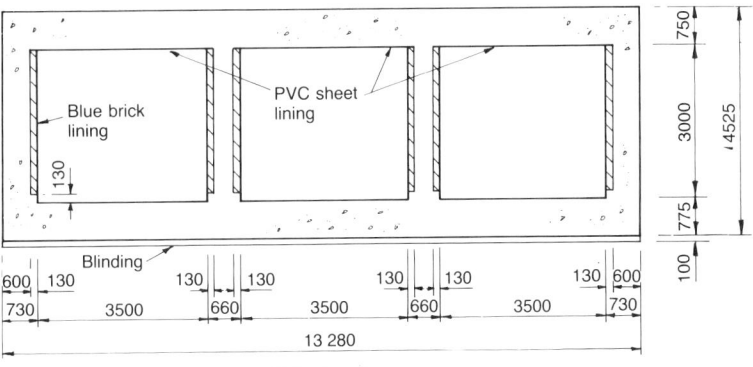

Culvert section

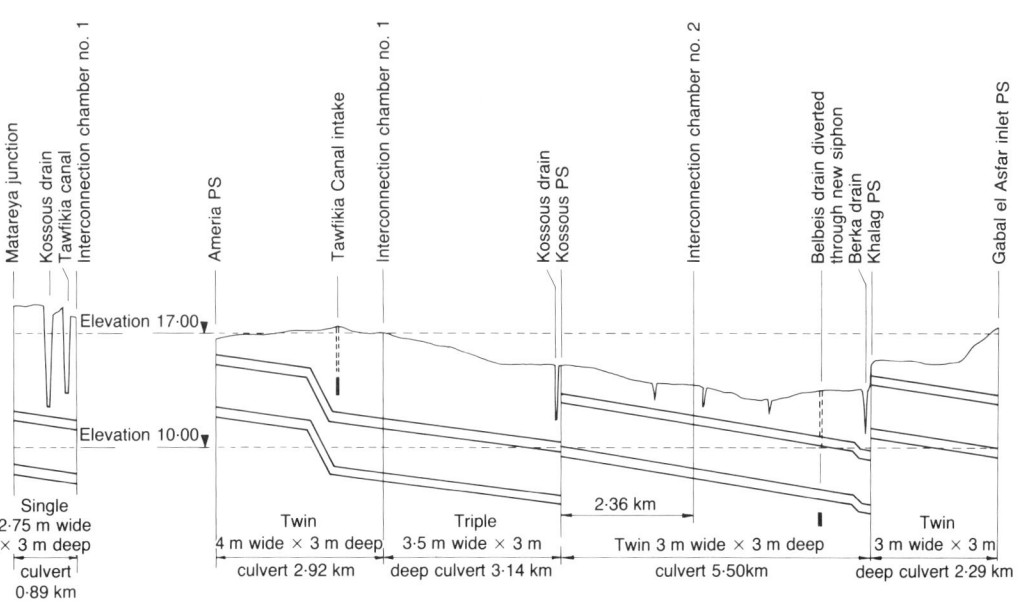

Culvert profile

Fig. 13. Conveyance system profile

is provided by acid-resistant bricks on the walls and PVC sheet on the roof soffit.

88.   Connections to the system are limited to the interconnection and reception chambers where pipe stubs fitted with penstocks are provided to facilitate future sewer connection. The depth of the culvert conveyor at each outfall chamber is fixed to provide a minimum cover of 1·2 m. The maximum depth of construction was 9·75 m and is consistent with the generally accepted maximum depth of economic conventional excavation in the water bearing soils.

Construction

89.   *Programme.*   The culvert conveyor was constructed under three contracts.

Contract 6   Culvert from Ameria PS to IC 1, IC 1 and Matareya branch culvert (The Arab Contractors, with Tarmac Overseas Ltd).

Contract 7   Culvert from IC 1 to Kossous PS (Société Egyptiene d'Entreprises with AMEC Group plc).

Contract 8   Culvert from Kossous PS to Khalag PS, IC 2 and culvert from Khalag PS to Gabal el Asfar WWTP (Construction and Reconstruction Co. with John Laing International Ltd).

Details of these contracts are given in fig. 5 of reference 1.

90.   *Diversions and site clearance.*   The site for culvert construction was a 40 m wide strip, except for the first 2 km of culvert from Ameria PS. Here the culvert followed the line of an irrigation canal and the site, including one lane of Ismailia Canal Street, was a minimum of 24 m wide.

91.   Most of the culvert conveyor was routed through rural areas, apart from the first 2 km from Ameria PS, the first 2 km from Kossous PS and the upstream part of the Matareya branch, which are located in developed areas. Construction of the culverts involved 11 major road, rail, canal and drain crossings. Several major water and petroleum pipelines crossed the culverts, as well as small irrigation canals, gravity sewers, cables, buried electricity and telephone cables and overhead electricity cables.

92.   Drains were temporarily diverted for construction of the culvert crossings. The Belbeis drain was also temporarily diverted to allow construction of the new siphon undercrossing, which was completed ahead of culvert construction. Temporary over-pumping was utilized for construction at canal crossings.

93.   *Excavation and groundwater control.* Construction of the culverts involved extensive excavation with a total of 1·2 million m³ of material removed under the three contracts. The soils along the conveyance system are alluvial and other marine deposits, consisting mostly of made ground overlying dense fine-grained soils, which in turn overlie sands and occasional gravels. The culverts were mostly founded in the fine-grained soils, silty clays

and dense fine sands. Over the last 1 km section of culvert towards Gabal el Asfar WWTP a clay layer intruded into the excavation. The water table was generally 2–3 m below ground level, except near Ameria, where the water table was 5–6 m deep, and near Gabal el Asfar, where the water table was almost at ground level.

94.   On contract 6, excavation for the first 2 km downstream of Ameria PS was carried out between heavy section sheet piles. Although this section was relatively shallow, sheet piles were required because of the proximity of four- and five-storey residential buildings on shallow foundations on one side and a heavily trafficked road, Ismailia Canal Street, and the Ismailia Canal on the other side. Dewatering for most of this section was accomplished using drainage channels and sump pumps. Downstream of the Tawfikia canal crossing the much deeper excavations were also carried out within sheet piles in order to keep construction activities within the 40 m wide site. Deep wells were used for dewatering along this section. Excavation was carried out by backacters from the surface, assisted by a bulldozer and loading shovel in the bottom of the excavation.

95.   The culvert constructed under contract 7 is located in a rural area, apart from a short length upstream of the Kossous PS. This is the largest of the culvert sections at 13·28 m wide and it is also the deepest, being founded at 9·75 m. In order to keep construction activities within the 40 m wide site the contractor elected to provide excavation support, but on one side only, with the other side battered and bermed. The excavation support comprised H-section piles at 1·5 m centres with timber infilling between the H-piles. Anchor piles were used to restrain the top of the piled support. Backacters and loading shovels were used for excavation and deep wells were used for dewatering. The 300 mm dia. deep wells on both sides of the excavation were staggered at 30 m centres.

96.   Like contract 7, the contract 8 culverts were mostly in agricultural land but the culverts were smaller in size, 8·12 m wide and at shallower depths for most of their length. Apart from a few short sections open cut battered excavation was possible within the 40 m wide site. Two-stage well point dewatering with well points at 1·0 m centres on each side of the excavation was used for almost all of the length of the culvert. Over the last 1 km towards Gabal el Asfar, where a deep clay seam was encountered, deep walls were employed in addition to the well point system.

97.   *Antiquities.*   During the design stage permission had been sought from the Antiquities Organization, which granted approvals, for the culvert conveyor routes. However, the finding of articles of antiquity should not be unexpected in Egypt.

98.   A first find was on contract 6 at the proposed batching of plant site. Permission was eventually granted for a modified site at the same location. Later serious disruption was caused by the discovery during site clearance for the upstream 300 m of the Matareya branch

culvert that the proposed culvert was located directly on the line of one of the walls of the ancient city of Heliopolis dating from Pharaonic times. A revised route was selected to ensure complete clearance between the culvert and the wall. Several months later a boat dating from Pharaonic times was uncovered in the side of the excavation on contract 7. Construction activities were delayed until the boat could be removed by the Antiquities Organization.

99. *Concrete works.* Shutter designs and concrete pours on all three culvert contracts were based on the 15 m long design modules. The modules were constructed on a hit and miss basis. Wall and roof shutters were prefabricated steel forms which were collapsible and retractable and were moved into place on rails. Base slabs were cast in one pour, with a maximum of 153 m^3, as were walls and roof slabs, with maximum pours 212 m^3 on contract 6 and contract 8. On contract 7 the roof slab was cast separately from the walls. The PVC sheet liner was laid on the roof shutter before fixing of the reinforcement and placing of the concrete.

100. Concrete was batched on site using 40 m^3/h capacity batching plants on contract 6 and contract 7, and transported in insulated trucks mixers. Concrete for contract 8 was batched off site in a 100 m^3/h capacity batching plant. Most of the concrete was placed by pump and at night, primarily because access was less congested and also because the temperatures were lower. Concrete curing methods varied, and included liquid curing membrane, hessian, PVC sheeting and ponding. Concrete cubes were tested on site on contracts 7 and 8 and in an off-site laboratory for contract 6. A total of 275 000 m^3 of concrete was placed on the three contracts.

101. *Watertight testing.* After the culvert had been backfilled and the dewatering system withdrawn, the culvert was visually inspected for any leaks which, if found, were sealed before the brick lining was placed.

102. *Brick lining.* All internal wall surfaces were lined with locally made acid-resistant blue bricks. The joints in the brick wall were pointed to a depth of 25 mm using an acid-resistant epoxy mortar. Joint preparation techniques included acid etching, wire brushing and taping of the bricks before laying in place. A variety of hand-operated pumps, compressed air driven pumps and specially shaped trowels was tested to determine the most suitable and efficient way to apply the mortar pointing. Hand application, although slower, incurred less wastage of the expensive mortar and avoided time-consuming cleaning of pumps and nozzles at the end of each shift. Bricks were tested for compliance with the specification at the rate of 4 every 10 000. The epoxy mortar was tested on site by making 25 mm dia. cylinders which were compression-tested at seven days, and the strengths were compared with control strengths measured during full-scale testing—which was carried out on each 20 t batch of mortar. A total of approximately 10 million bricks was used to line the culvert walls.

Problems

103. All three contracts suffered from delayed possession of the site or parts of the site, inadequate supplies of blue bricks, shortages of cement and delays in payment. One particular farm on contract 8 was the subject of several court cases until acquisition was completed in 1992. Contract 6 suffered additional delays as a result of the time needed to complete preparations for extensive traffic diversions, and also because of the discovery of antiquities. The actual construction periods for each contract were

contract 6   64 months
contract 7   58 months
contract 8   86 months

Commissioning

104. The first sewage flows were diverted into the new culvert conveyor on 13 January 1992. The sewage flows were discharged to the Kossous drain at Kossous PS, which serves as the disposal outlet until the Gabal el Asfar Inlet PS and bypass channel have been completed. Flows are restricted to one box of the culverts in order to ensure that the minimum design velocities are achieved at the initial flow rates.

References

1. TAYLOR BINNIE & PARTNERS. *Wastewater master plan report*, TBP, London.
2. KELL, A. D. K. *et al.* Project objectives, organization and implementation. *Proc. Instn Civ. Engrs, Civ. Engng, Greater Cairo Wastewater Project*, 1993 8–17.
3. O'KANE and REECE. Post-construction services. *Proc. Instn Civ. Engrs, Civ. Engng, Greater Cairo Wastewater Project*, 1993 64–67.

*Proc. Instn
Civ Engrs,
Civ. Engng,
Greater Cairo
Wastewater
Project,
1993, 48–55*

Paper 10230

Gabal el Asfar treatment plant

I. Taylor, BSc, MSc, DIC, MICE, *Eng. M. Halim*, BSc, *and
M. Wishart*, DipCivilEng, MICE, MIWEM

*I. P. Taylor,
Binnie &
Partners;
Chief Resident
Engineer — WWTP,
AMBRIC*

*Eng. M. Halim,
Project Resident
Engineer,
CWO — WWTP
and Member of
CWO Board of
Management*

*M. Wishart,
Acer Consultants
Limited;
Team Leader
Design
Review — WWTP,
AMBRIC*

■ Principal design criteria

The master plan and further studies determined that the capacity of the plant will ultimately be 3×10^6 m^3/day to serve 12 million people and that it will be built in stages with the first stage being for 10^6 m^3/day. The studies also determined plant influent parameters of 335 mg/l of BOD and 600 mg/l SS, wastewater principally of domestic origin with a high grit content. A stage 1 average flow of 10^6 m^3/day was determined with a peak diurnal flow of 1.5×10^6 m^3/day.

2. To meet the Egyptian effluent quality legal requirements (outlined in Table 1), the design standards for the plant were determined to deliver a 30/30 BOD/SS standard effluent at peak diurnal flow. Chlorination facilities are required to satisfy the 0·5 mg/l residual requirement for discharge to drain under all flow conditions. The design effluent quality is not suitable for unrestricted use in irrigation but Egyptian law does permit the use of lower quality waters for some purposes, e.g. desert reclamation.

3. The sludge treatment facilities are designed to deliver a dewatered digested sludge with a minimum solids content of 30% by weight.

4. The Contractor was required to guarantee the performance of both the wastewater and sludge treatment facilities under given influent loadings and within guaranteed operational consumption figures for electricity, chlorine, polyelectrolyte and diesel fuel oil. He was further required to guarantee a minimum generated electricity output for the dual fuel generating plant operating on gas produced by the sludge digestion process. In the event of non-compliance with the guaranteed consumption levels a proportional refund formula may be implemented.

Site and treatment plant outline

5. The site for the stage 1 plant works is approximately 2 km long and 1 km wide. The layout of the treatment facilities has been arranged with flow passing west to east, to discharge treated effluent to the Gabal el Asfar drain which has been realigned to bisect the site (Fig. 1). The drain itself discharges to the Belbeis drain and ultimately to Lake Manzala on the Mediterranean coast.

6. The plant is of the conventional non-nitrifying activated sludge type and includes screening and grit removal, primary sedimentation, surface aeration, final clarification and chlorination facilities. The selection of this type of plant was made taking into account land availability, full life cost, robustness of construction, ease of operation, maintenance and environmental considerations. The sludge treatment facilities comprise thickening in two stages: anaerobic digestion and mechanical dewatering. Sludge gas in excess of demand for digester heating will be used for power generation with surplus automatically wasted to a flare stack.

7. The stage 1 wastewater plant is arranged in four modules, each providing treatment for 0.25×10^6 m^3/day with common ancillary service buildings; the sludge digestion facilities are arranged in two modules with common mechanical dewatering, power generation and ancillary service buildings. On completion approximately 75% of the assigned area will be fully developed with process plant and the remainder left available for short-term stacking of dewatered sludge. There is some demand for dried sludge as a soil conditioner but most of it will be disposed of to desert landfill sites currently being determined by the Cairo Wastewater Organization.

Table 1. Egyptian effluent quality legal requirements

	Limiting standards	
	Discharge to drains: mg/l	Reuse for agriculture*: mg/l
BOD	40	250
SS	40	50
Residual chlorine (20 min)	0·5	—
Coliform count per 100 ml	—	100

* Depending upon the type of crop and the method of irrigation employed.

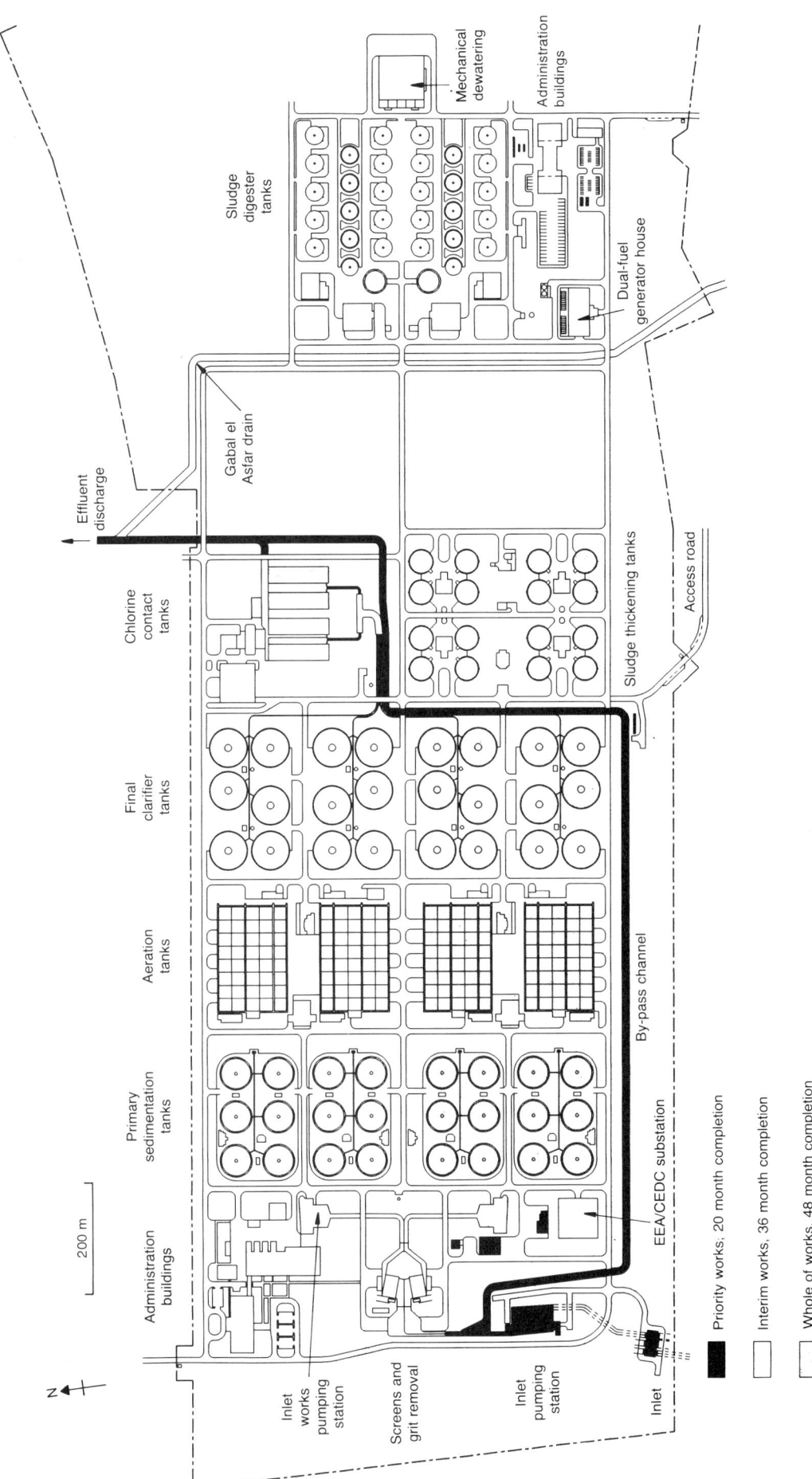

Fig. 1. Site layout

Table 2. Summary of principal facilities*

Structure unit/item of plant	Equipment	Number	Unit size, design parameters and ratings
Inlet pumping station	Archimedean screw pumps Steel troughs and glass-reinforced plastic covers	10	3·15 m dia., 9·6 m lift, 38° inclination, 2500 l/s capacity Motor 3·15 kV, 500 kW, 1500 rev/min
Works bypass channel	Flow to works control penstocks Bypass control penstocks	2 2	2500 mm × 2200 mm channel-mounted motor-driven 3000 mm × 2200 mm channel-mounted motor-driven
Primary distribution substation	Transformers Medium and low voltage switchboards	4	10·5–3·15 kV, 10·5 kV–380 V transformers 10·5 kV, 3·15 kV, 380 V switchgear
On-site generator house	Diesel-driven generators	3	Diesel V16 engine 750 rev/min 271·83 litres Alternator 10·5 kV, 2320 kW
Screens unit (2)	Coarse, hand-raked inclined straight bar Medium, mechanically-raked straight bar	14 14	1·6 m wide, 2·9 m deep, 65 mm bar spacing 2·4 m wide, 2·9 m deep, 20 mm bar spacing
Grit removal unit (2)	Rectangular diffused air horizontal flow, scraper air lift, reciprocating elevator rake Rotary blowers	8 10	2 channels/tank 18 m long, 3·6 m wide Retention time 3 min, forward velocity 0·1 m/s, 220 t/day grit yield 810 m³/h at 500 mbar, 18 kW
Inlet works pump station (2)	Archimedean screw pumps Steel troughs and glass-reinforced plastic covers	10	3·15 m dia., 9·2 m lift, 38° inclination, 2550 l/s capacity Motor 3·15 kV, 500 kW, 1500 rev/min
Primary sedimentation tanks	Circular, rotating bridge 0·6 m dia. scraper, timer or manual desludging	24	Radial flow tanks 42 m dia., 3·0 m side water depth Central sludge hopper 6·0 m dia., flow 2112 m³/h, retention time 2·25 h
Primary sludge pumping station (4)	Centrifugal vertical spindle pumps Progressive cavity pumps	16 8	Centrifugal pumps 127 l/s at 37 m head Progressive cavity pumps 22 l/s at 60 m head
Aeration tanks (4 streams)	Plug flow channels Surface aerators	20 160	Each channel 140 m long, 17 m wide, 8 cells per channel 2·8 h retention time, Simcar SAM aerators, 75 kW
Final clarifier	Circular, rotating bridge 0·6 dia. chevron scraper, hydrostatic continuous desludging	24	Radial flow tanks 53 m dia., 3·1 m side-water depth Hydrostatic sludge lift system, flow 3390 m³/h, retention time 2·5 h
Return-activated sludge pumping station (2)	Archimedean screw pumps	10	2·6 m dia., 7·85 m lift, 38° inclination, 1450 l/s capacity Motor 380 V, 250 kW, 1500 rev/min
Waste-activated sludge pumping station (2)	Centrifugal vertical spindle pumps (two-speed)	12	120 l/s at 10·5–19·3 m head
Chlorine contact tank	Injectors, residual monitoring	4	Contact tanks 80 m long, 26 m wide, 2·5 m water depth

Table 2—continued

Structure unit/item of plant	Equipment	Number	Unit size, design parameters and ratings
Chlorine house	Chlorine drum store/evaporators	120/12	Drum nominal capacity 1000 kg each
	Chlorinators/injectors/motive pumps	12/12/12	Maximum dosage rate 266 kg/h (15 mg/l)
Chlorinated effluent pumping station (2)	Centrifugal vertical spindle effluent pumps	8	350 l/s at 16 m head
	Centrifugal vertical spindle irrigation pumps	6	50 l/s at 43 m head
	Hydro-pneumatic booster sets	2	One 105 l/s at 8 bar pressure, one 195 l/s at 6 bar pressure
Sludge blending and dilution chambers	Picket fence agitator		Blending of primary and waste-activated sludges and dilution/conditioning of mixed sludge with chlorinated effluent
Sludge thickening tank	Circular, hopper-bottomed, fixed bridge picket fence thickener and echelon scraper	16	32 m dia., 4.0 m side wall depth, full fixed bridge picket fence, centre torque slewing ring drive, central sludge draw-off
Thickened sludge pump station (4)	Positive displacement ram pumps	16	Double-acting ram pumps 25 l/s at 58.5 m head
Digester feed pump house (2) (includes drum thickeners)	Centrifugal pumps/drum thickeners	10/10	Pumps 1.95–9.72 l/s at 60 m head drum thickeners 88.6 m³/h
	Polyelectrolyte dosing/progressive cavity pumps	2/30	Pumps 26.61 l/s at 12 m head
Primary digester (20)	Heating and mixing circulators	180	Digesters 30.0 m i.d., 15.27 m sludge depth Heaters/circulators 11.38 m long, 100 000 kcal/h heat output
Secondary digester	Adjustable bellmouth draw-off	10	Digesters 26.5 m i.d., 13.7 m sludge depth Bellmouth 300 mm dia., 1000 mm adjustment range
Compressor and boiler house (2)	Rotary compressors	26	Sludge gas compressors 1220 m³/h at 1.8 bar
	Dual-fuel boilers	10	Boilers 1 600 000 kcal/h net capacity each
Gas holder	Telescopic double lift gas holder	2	30 m i.d., 11 000 m³ capacity
Mechanical dewatering house	Progressive cavity pumps/polyelectrolyte dosing	30/2	Pumps 1.94–8.88 l/s at 15 m head
	Belt presses	30	Belt press flow rate 2.5 m³/h, dry solid load 990 kg/h
Dual-fuel generator house	Dual-fuel driven generators	10	Sludge gas–diesel V16 engine 750 rev/min, 271.83 litres Alternator 10.5 kV, 2320 kW
Substations (17)	Transformers	49	10.5 kV–380 V transformers
	Medium and low voltage switchboards		10.5 kV and 380 V switchgear

* Loading/criteria are based on average flows and pump ratings on maximum output; equipment numbers include standby/maintenance units.

Hydraulics through the wastewater treatment plant

8. Figure 2 shows the treatment plant hydraulic profile as finally adopted at maximum water levels. The tender design incorporated two intermediate lift-pumping stations: one immediately upstream of the primary sedimentation units, and a second to lift final clarifier effluent to the chlorine contact tanks. The Contractor elected to redesign the profile to maximize the lift of the inlet works pumping station, elevating the hydraulic profile through the works and eliminating the effluent lift station. This change also serves to minimize the construction difficulties and costs associated with the extensive dewatering needed by the original profile, at the expense of an additional above ground cost.

Wastewater and sludge treatment facilities

9. Table 2 summarizes the principal facilities of the wastewater treatment plant.

10. Raw sewage reaches the plant at a flow reception and distribution chamber designed to receive and balance flows for all three stages (3×10^6 m^3/day) of the treatment plant development. A stage 1 dedicated screw pump station lifts the flow to feed two screening and grit removal units. Each screening unit comprises seven screening channels incorporating coarse hand-raked and medium mechanically-raked straight bar screens. Conveyors discharge screenings to containers for off-site disposal. Each grit removal unit comprises four aerated grit channels with continuous grit removal by scraper, air lift and reciprocating elevator rake discharging to a grit bunker. Wet screenings yield for the two units is anticipated to be 40 m^3/day. Maximum grit yield is estimated at 220 t/day.

11. After measurement, flow passes to two intermediate lift screw pumping stations, where it is again divided to feed two 0.25×10^6 m^3/day primary and secondary treatment modules.

12. Six 42 m dia. primary sedimentation tanks per module are equipped with spiral configuration, peripherally driven rotating bridge scrapers. Tanks have 3 m side water depth and 7.5° floor slope. Settled sewage discharges to inboard launders via two peripheral V-notch weirs. Hydrostatic desludging at approximately 1% solids content is under timer or manual control. Tanks are constructed on up to 4 m depth of fill with the central hopper extending into natural ground.

13. Aeration tanks in each module comprise five parallel aeration channels 140 m long, 17 m wide and 4.5 m deep, with eight cells per channel, each equipped with bridge-hung surface aerators and cruciform anti-vortex baffles. Adjustable outlet weirs are provided to vary aerator immersion to give dissolved oxygen control. Peak total oxygen demand for the plant is 320 t/day. The tanks are equipped with travelling cranes for aerator movement and are founded on 1 m of filled ground.

14. Six upward flow 53 m dia. final clarifiers per module discharge effluent by V-notch weirs to an outboard peripheral channel and an inboard launder 36 m in diameter. The tanks, founded in natural ground, have a floor slope of 4.8° between the side wall and the intermediate launder supports, and are almost flat-bottomed inboard. Continuous, manually adjustable, sludge removal is by a rotating bridge chevron scraper, with six hydrostatic lift pipes per unit sweeping the flat-bottomed areas of the tank.

15. Final clarifier effluent from all modules is blended and distributed to four chlorine contact tanks. The single chlorination house accommodates four chlorination units which each have two pairs of drums, three evaporators and three chlorinators. Motive water is drawn from the chlorinated effluent channel. Up to 250 kg/h of chlorine will be required to maintain the residual of 0.5 mg/l at 30 min. The chlorinated effluent finally discharges to the Gabal el Asfar drain.

16. A full flow plant bypass channel is provided. Under emergency conditions the 1750 m long, 9 m wide channel can divert untreated sewage via an overflow weir and penstocks downstream of the inlet pumping station to discharge direct to the outfall and Gabal el Asfar drain.

17. Activated sludge is recirculated through measuring flumes and via two screw-pumping stations which each serve a pair of flow modules. Surplus activated sludge is forwarded by centrifugal pumps to a blending chamber and mixed with primary sludge forwarded by the four primary sludge pumping stations. Primary sludge pumping is by centrifugal pumps, but stations include provision of progressive cavity pumps for high sludge solids content events and blockage clearance.

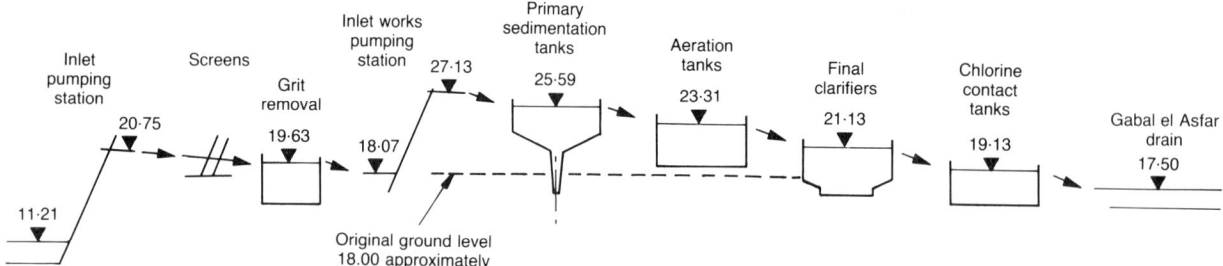

Fig. 2. Hydraulic profile: maximum water levels in metres

18. Blended sludge is conditioned by the addition of chlorinated effluent to maintain a solids content 3500–5000 mg/l for feed to four modules of four 32 m dia. gravity thickening tanks. Tanks are equipped with fixed bridge picket fence thickeners and scrapers to discharge sludge, thickened from approximately 1% to a minimum of 4% solids content, to a central hopper for hydrostatic draw-off. Liquors are returned by gravity to the flow reception chamber at the head of the plant.

19. Thickened sludge is forwarded, approximately 650 m, by double-acting ram pumps to the sludge treatment facilities located to the east of the Gabal el Asfar drain.

20. Before it is forwarded for digestion, the combined thickened siudge is further thickened to a minimum of 6% solids content by centrifugal drum thickeners in the two digester feed pumping stations. Sludge is fed to the thickeners by centrifugal pump with polyelectrolyte injection on the delivery lines. Anticipated polyelectrolyte consumption is 920 kg/day.

21. Finally, thickened sludge is forwarded by individually dedicated progressive cavity pumps to ten primary digesters per treatment module. Designed for a minimum of 20 days' retention, the digesters are of fixed (concrete) roof construction with sludge gas displaced to independent two-stage telescopic floating bell gas holders. Tank contents are heated and mixed by roof-hung heating circulators to maintain a design operating temperature of 35°C and to achieve complete and uniform mixing of fresh feedstock within 40 min of intermittent tank charging. Circulation is by recirculated compressed sludge gas and heating by water-sludge heat exchanger, incorporated as a jacket to the circulator, on a low-pressure closed-loop hot water system with dual-fuel-fired boilers (sludge gas and diesel). Gas compressors, blowers and boilers are housed in an integral compressor and boiler house serving each treatment module.

22. Digested sludge gravitates to five open-topped, hopper-bottomed secondary digesters per treatment module. Secondary digesters are designed with seven days' retention capacity. Decanted liquors discharge to a liquor pumping station and are forwarded to the head of the liquor drain in the wastewater plant for return to the head of the plant.

23. Digested sludge from each digestion module gravitates to a pair of sludge holding tanks adjoining the mechanical dewatering house. The holding tanks are equipped with paddle agitators to maintain uniform solids suspension, and act as wet wells for the dewatering machine feed pumps. Dewatering is by belt filter presses with dedicated progressive cavity feed pumps and dedicated coagulant dosing arrangements. Polyelectrolyte dosing is automatically controlled against sludge flow and solids content; a consumption of 800–1100 kg/day is estimated. Sludge cake discharges to a conveyor system which delivers to a discharge station for trucking to the sludge stacking area. Liquors gravitate to the liquor pumping station. The plant is designed for duty equipment to dewater one week's sludge production in six-day 24 h operation with an estimated 525 t dry solids cake at 30% solids content product per working day.

Electrical power

24. The estimated peak electrical power demand for the plant is 26·8 MW. The demand will be met from two sources: the generation/distribution authority grid, through a 63/10·5 kV substation in the wastewater plant, and dual-fuel generation plant in the sludge treatment area. Although rated to meet the full demand, the grid source is intended primarily to supplement the output from the dual-fuel plant. The sludge gas generated is estimated to be sufficient to supply approximately 18·3 MW continuously.

25. Interconnected double busbar switchgear at the primary distribution substation in the wastewater plant and the dual-fuel generator house substation permits the supply at any load centre to be sourced from either the grid or the plant-generated supply. Under normal operating conditions the grid will supply the inlet pumping station, the wastewater plant administration area and 50% of the aeration plant load. Because of the strategic importance of the inlet pumping station to the upstream conveyance system, standby diesel generators are installed to keep this pumping station operational on grid supply failure.

Control philosophy and provisions

26. The control philosophy adopted generally relies on the manual control of plant and equipment at local control centres. The system is structured in three tiers, with local control centres under the supervision of process-related area monitoring centres, where data are marshalled and displayed, which report via a high-speed data link to the main monitoring centre in the wastewater plant administration building.

27. Automatic control systems incorporated within the plant process controls include emergency shutdown of inlet pumping station pumps, automatic (timer) desludging of primary tanks, control of aeration via dissolved oxygen monitoring, control of digester feed pumps and gas recirculation compressors, and polyelectrolyte dosing systems. All pumping stations, except the screw-pump stations, are under automatic level control.

28. To assist in the operation of the dual-fuel generating plant, a comprehensive energy management system is provided, giving details on gas production, storage and quality, together with actual electrical power loadings and their distribution against grid or works generated supply.

Foundation and structural design

29. The site is within the area of the existing Gabal el Asfar sewage farm. It is very flat, criss-crossed by abandoned irrigation canals and drains and has a groundwater table gener-

ally 1·0–1·5 m below ground level. The upper stratum of top soil and silty clay is 0·2–1·2 m thick and overlies a layer of silty sand 4–8 m thick. This is underlain by a stiff to very stiff clay 1–12 m thick, over a very dense sand.

30. With the exception of the screw-pumping stations, which required piled foundations, all process and liquid conveyance structures have raft foundations; a rigid raft design was adopted for the smaller and more deeply founded structures, and a flexible raft type with movement joints was used for the aeration tanks, final clarifiers and chlorine contact tanks to accommodate moderate differential settlements. Structural designs complying with CP 110 and BS 5337 were adopted.[1, 2] Grade 25 concrete was used throughout and the reinforcing steel was mainly grade 460.

Operations and maintenance provisions

31. Overall management of the plant is centralized in the administration and laboratory building in the wastewater treatment plant.

32. To provide for uninterrupted full operation of the plant, strategic standby and maintenance provisions are included. Fully equipped standby screening and grit removal channels are provided. Hydraulic designs allow for any one primary tank, a pair of aeration channels and a final clarifier per module, and one chlorine contact tank to be decommissioned for maintenance. An installed standby/maintenance equipment provision of 20–25% is made for process flow pumping and sludge dewatering equipment.

Contract details

33. Although tendered as a single remeasurable contract, before it was awarded the contract was split, principally for financial reasons, into two complementary lump sum contracts—contract 16·1 and contract 16·2—with the Contractors responsible for the detailed design and performance of the plant. The wastewater plant design is based on AMBRIC's detailed design, but incorporates accepted changes to the hydraulic profile, the grit removal facilities and the number and size of the sedimentation tanks and final clarifiers. The sludge treatment and power generation plant design fully develops the accepted tender proposal which was based on a performance specification. Contract 16·1 covers the design, construction and commissioning of the sludge treatment facilities to the east of the Gabal el Asfar drain, together with the flow reception chamber and the inlet pumping station at the head of the plant. Contract 16·2 covers the main wastewater treatment plant and the first stage sludge-thickening facilities.

34. The firms making up the two joint ventures undertaking the works are ANSCO (Ansaldo SpA with Condotte, Italy) as joint venture leader for both contracts, FOCHI (Filippo Fochi SpA, Italy) on both contracts, with UNION (Union of Public Sector Organisation for Construction Companies of Egypt) on

contract 16·1 and with Nasr General Contracting Company of Egypt on contract 16·2.

35. Finance for the two contracts is from the Egyptian and Italian governments, the Egyptian pounds component being funded completely locally and the US dollar component being funded by the Italian government through soft loans (45%), buyer credits (52%) and a grant (3%). The US dollar component of the financing was arranged by the Italian joint venture members as part of their Tender.

36. Lump sum contracts were awarded to the two joint ventures in June 1990 and July 1990 and came into force about 18 months later on disbursement of agreed local and foreign currency advance payments. Contract 16·1 was for US$ 88·2 million and contract 16·2 for US$ 172·8 million.

37. The Contract requirements for phased completion are shown in Fig. 1. Priority works, to be completed in 20 months, include the flow reception chamber, inlet pumping station, plant bypass channel and related ancillaries, including the primary distribution substation and on-site generator house. The two northern treatment modules of the wastewater plant, 50% of the chlorination and first stage sludge thickening and the full sludge treatment facilities to the east of the Gabal el Asfar drain are to be completed within 36 months. The remainder of the treatment plant has to be completed within 48 months. In addition the joint ventures are required to operate and maintain the works for 12 months and to provide technical training to the staff of Cairo Wastewater Organization and GOSD

Design check and site supervision

38. AMBRIC, in association with Moharram–Bakhoum (ACE), was appointed as the Engineer for both contracts in November 1991. Design check and factory inspection services are jointly funded by the Egyptian and British governments (the latter through ODA grants). Site supervision services were funded entirely by the Government of Egypt. The construction supervision teams were set up so that the Egyptian members of the team carry out the major technical and administrative role; the expatriates provide management and advisory input. The CWO maintain a major office at the site in order to follow closely the execution of the works.

Principal resource requirements and sources

39. Wherever possible, locally produced materials are used; imported materials are limited to those specified or where suitable alternatives are not readily available. For example, the use of large-diameter prestressed concrete pipes has been permitted in place of the specified ductile iron pipes in some locations.

40. More than 70 km of major pipelines are required, excluding irrigation, washwater, potable water services and so on, as shown in Table 3.

Table 3. Materials of major pipelines

Material	Diameter range: mm	Length: m
Ductile iron	80–1000	29 500
Carbon steel	50–600	19 500
Reinforced concrete	800–2000	6 000
Vitrified clay	150–750	15 250

41. To complete the works the joint ventures will place over of 520 000 m³ of grade 25 concrete with a forecast peak production rate of 30 000 m³ per month.

42. Specialized mechanical and electrical equipment will come predominantly from Italy, but there is a requirement for Egyptian materials, accessories and fabrication facilities to be used wherever possible. All cables and medium and low voltage switchboards are of local manufacture and assembly. Bridge scrapers, picket thickeners and gas holders are typical of the equipment which is partly imported and partly assembled locally.

Construction works

43. The construction method is generally a conventional labour-intensive one. In January 1993 more than 1500 contractors' staff were working on the project. Piling for the inlet and inlet works screw-pump stations has been completed with 100 t and 115 t cast in situ reinforced concrete piles, 18–23 m long and 650 mm in diameter, founded in the dense sand underlying the stiff clay. For the return activated sludge pumping stations, 18 m long and 60 t precast driven piles were used.

44. The management of groundwater disposal from multiple sources to the available irrigation channels and soakaways demands careful co-ordination and programming of excavation activities. For deep excavations (e.g. the flow reception chamber), contractors have used deep wells and sand drains combined with battered excavations. For shallower structures (e.g. primary tanks), combinations of shallow wells and sump pumps are used, sometimes in combination with sheet pile cofferdams.

45. Custom-designed shutters are used for repetitive wall construction as on the primary and final tanks, and slip-form construction has been adopted for primary and secondary digester walls. Two methods of primary digester roof placement are used; both include casting the roof at base slab level, with lifting/jacking to position with the slip-form, and hoisting to place with jacks reacting on completed walls.

46. To gain time contractors are employing precast solutions where practicable, notably for roof beams and slabs to the superstructures of the priority works' and for primary and final tank weir channels and launders where casting can proceed in parallel with tank construction.

Conclusion

47. The Gabal el Asfar stage 1 plant is one of the largest ever built in one construction project. When completed it will ensure that most of the wastewater generated on the East Bank is treated to acceptable standards, and it will enable large quantities of treated effluent and dried sewage sludge to be available for reuse in agriculture.

References

1. BRITISH STANDARDS INSTITUTION. *The structural use of concrete*. BSI, London, 1972, CP 110.
2. BRITISH STANDARDS INSTITUTION. *Code of practice for the structural use of concrete for retaining aqueous liquids*. BSI, London 1976, BS 5337.

Proc. Instn
Civ. Engrs,
Civ. Engng,
Greater Cairo
Wastewater
Project,
1993, 56–59

Paper 10231

West Bank scheme

*G. R. Miller, BSc, MSc, PhD, MWEF,
MAGU, and R. J. Kachinsky, BS, MS,
DEE, FASCE*

G. R. Miller,
Black & Veatch
International;
Project Director,
AMBRIC

■ Planning

Concurrent with the rehabilitation and construction of the East Bank facilities, work has progressed on the West Bank of Cairo, funded principally by grants from the United States Agency for International Development (USAID) and by the Government of Egypt. The original strategy was set up in the late 1970s and is still in place; however, the original programme has slipped since the strategy was devised.

2. Generally, planning was based on the following strategy: as a temporary step, remove wastewater from the streets by rehabilitating the existing system; devise methods to collect wastewater from the large and ever-growing number of unsewered areas; design and build a new, larger collection and transport system and connect it to the existing system; build treatment plants as rapidly as financing will allow; train local staff to administer, operate and maintain the system; expand the collection system into the unsewered areas in conjunction with the water system expansion.

Rehabilitation

3. As the master plan and new works programme were developed, it was also necessary to develop a rehabilitation programme which would maintain and improve the efficiency and reliability of the existing works until the new works could be commissioned. Preliminary studies indicated that rehabilitation of the subsidiary pumping station, ejectors and major pumping stations would be most effective in meeting this objective. There would also be considerable benefit if the treatment efficiency of the existing wastewater treatment plants (WWTPs) could be improved.

4. Of the five existing treatment plants in Cairo, only the West Bank's Zenein WWTP was considered capable of improvement. It was in fair structural but poor mechanical and electrical condition. In addition, process and hydraulic problems existed. When it was commissioned in 1970, the Zenein wastewater treatment plant was a 0·22 million m³/day activated sludge treatment plant. An expansion of the plant to 0·33 million m³/day began five years later and was finished in 1985.

5. Because of overloading, design inadequacies, and improper operation and maintenance, the Zenein plant never performed as an effective secondary treatment plant. Flow is discharged into the Nahya drain and eventually reaches the River Nile. AMBRIC's final rehabili-

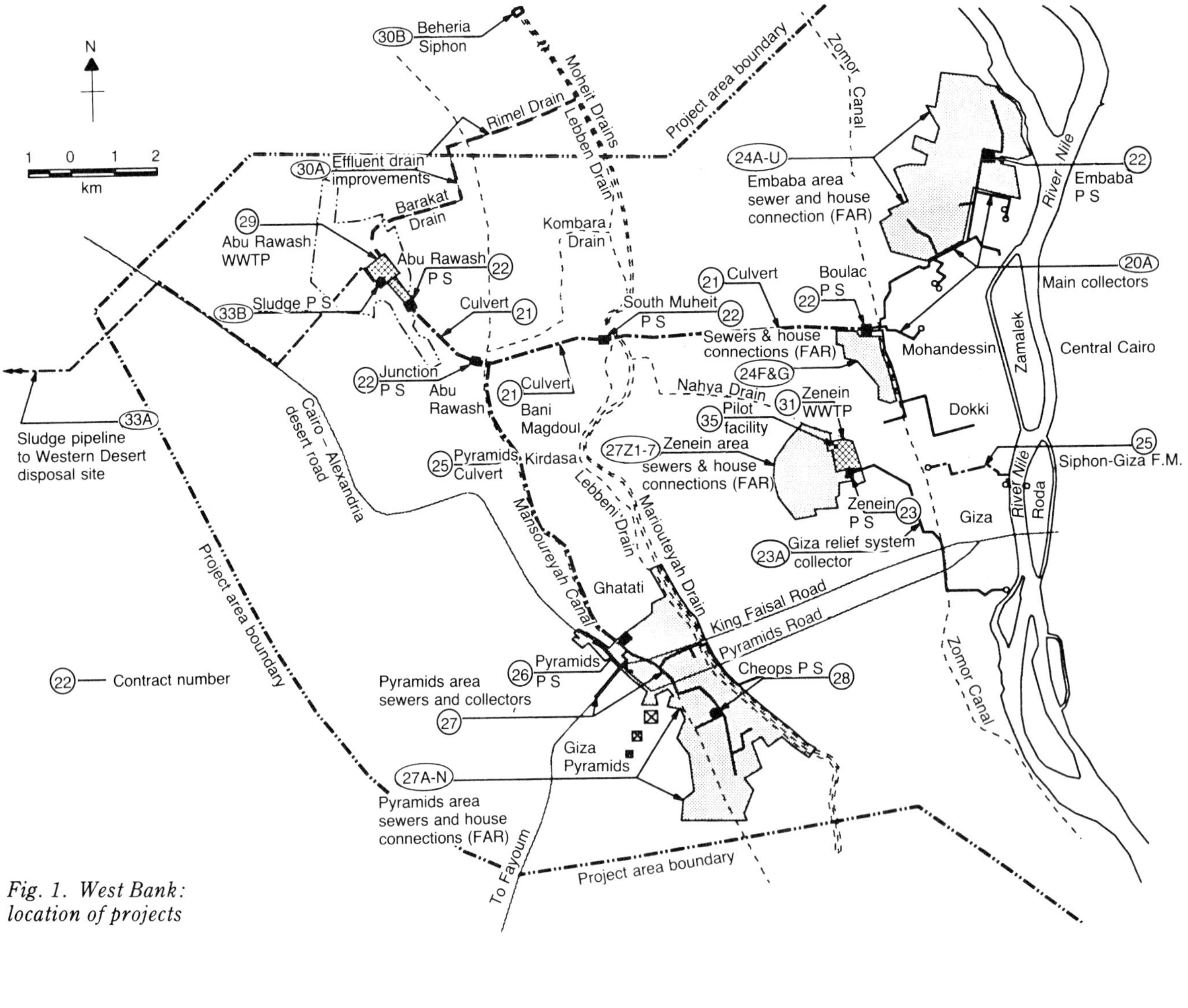

*Fig. 1. West Bank:
location of projects*

tation design included general mechanical, electrical and structural renovation/replacement for various process facilities and the addition of several new facilities. The most significant modification was the conversion of several aeration basins to biofilters to produce a combined biofilter/activated sludge process. Rapid sludge return systems and a chlorination system were also added.

Design

6. Major conveyance systems were proposed to carry existing and future wastewater to the rehabilitated WWTP at Zenein and the new wastewater treatment plant at Abu Rawash (Fig. 1).

7. West Bank unsewered areas were to be sewered. Therefore, new large diameter sewers were proposed to connect existing and newly sewered areas to a series of box culverts and screw pumping stations. The major collector system in the Embaba area is 18 km in length, has pipe diameters from 800 to 2750 mm, and carries wastewater from Embaba, Mohandesein and Dokki to the Boulac pumping station. The major new collector to the Zenein WWTP is 5·5 km in length, has pipe diameters from 1200 to 2500 mm and carries wastewater from Giza to Zenein. The construction of a 12·4 km collector in the Pyramids area is nearing completion. Data on these and other West Bank construction contracts are given in Table 1.

8. For pipework up to 600 mm in diameter, extra-strength vitrified clay pipe with flexible compression joints, manufactured in Egypt, was used. Reinforced concrete pipes were used for diameters of 600 mm or larger. Internal PVC linings were installed to protect the upper 270° of the concrete pipe against hydrogen sulphide corrosion.

9. Culverts were of cast-in-place concrete construction. To protect against corrosion, the walls were lined with blue bricks and pointed with acid-resistant mortar. Keyed PVC membrane sheeting was applied on the soffits. Resistant iron sluice gates were used for flow control at the isolation, interconnection and access chambers.

10. On the West Bank the north-west culvert serves the north and central developed area near the river, and the Pyramids culvert serves the developing areas around the Giza Pyramids. Boulac and South Muheit pumping stations lift the north-west culvert flows *en route* to the Junction pumping station where flows from the Pyramids and north-west culverts are pumped to the Abu Rawash WWTP.

11. The West Bank scheme includes an extensive programme to construct secondary and lateral sewers and connect homes and commercial buildings within unsewered areas to the new system. Much of the construction is in areas where standard excavation equipment is not effective and extensive local labour is required. For all these reasons, a programme was developed whereby the work would be executed by Egyptian contractors qualified to perform the construction required.

12. The programme is based on USAID grant assistance through a fixed amount reimbursable (FAR) agreement. The FAR agreements, executed between the Government of Egypt (through CWO) and USAID, are based on the following major policies.

(a) The FAR programme will provide sewers and service connections in unsewered areas.

(b) It provides a steady flow of work to Egyptian contractors with the intent of strengthening their abilities to manage and construct sewage facilities to appropriate standards.

(c) It uses the special capabilities of Egyptian contractors to work in narrow streets and confined conditions.

(d) FAR work is 79% funded by USAID through the conversion of US$ to £E, with the Egyptian government funding the balance.

(e) The method of payment encourages completion of the work with a minimum of disruption to local residents.

R. J. Kachinsky, Camp Dresser & McKee International Inc.; West Bank Manager, AMBRIC

Table 1. Physical data on large West Bank construction contracts

Sewers and culverts

Contract	Type	Barrels (number)	Dimensions		
			Size per barrel	Length: km	Capacity: m³/day × 10³
20A	Circular	1	0·8–2·75 m dia.	18·0	620
21	Box	2	3·0 m W × 2·0 m H	10·4	650
23A	Circular	1	1·2–2·5 m dia.	5·5	393
25	Box	2	2·6 m W × 2·0 m H	9·0	342
27	Circular	1	0·8–2·75 m dia.	12·4	684

Lift stations

Contract	Name	Screws (number installed)	Lift: m	Capacity: m³/day × 10³
22	Embaba	3	6·1	240
	Boulac	4	8·5	624
	South Muheit	4	8·0	624
	Junction	5	7·1	832
	Abu Rawash	5	6·5	832
23	Zenein	4 × 2	8·5 + 7·5	393
26	Pyramids	3	8·5	416
28	Cheops	2	5·9	100

Treatment plants

Contract	Name	Type	Capacity: m³/day × 10³
29	Abu Rawash	Primary	400
31	Zenein	Secondary	330

Drains

Contract	Name	Length: m	Capacity: m³/day × 10³
30	Improvements to Barakat, Abdel Rahman and Rimel drains	4350	1600
	Improvements to Beheria Canal siphon	175	2000

Sludge disposal facilities

33A	Force main and distribution system	33625	—
33B	Sludge pump stations	—	34
35	Sludge digestion/dewatering pilot plant	—	—

Table 2. West Bank laterals and service connections (FAR agreements)

Location	Contract No.	Contract status				Estimated value*: £E × 10⁶	Population served
		Complete	Under construction	To be constructed	Total		
Embaba	24A-T	7	11	3	21	186·1	740 000
Pyramids	27A-N	5	2	8	15	104·5	460 500
Saft El Laban	27Z1-7	2	5		7	52·2	179 700
Abu Rawah	34A-B			2	2		
Ghatati	34D			1	1		
Kirdasa	34E-K			7	7		
Kirdasa Pumping Station	34C		1		1		
Beni Magdoul	34L-M			2	2	110·0	430 00
Totals		14	18	24	56	452·8	1 810 600

* US$1·0 = £E3·3, March 1993.

(*f*) The work is divided between public and private sector Egyptian contractors, with the majority going to the private sector.

13. Design activities under the FAR programme started in July 1988, and the first construction contract was awarded in November 1988. When the 56 FAR contracts have been completed, 475 km of sewers, 20 300 manholes, 74 000 connection chambers and 561 000 service connections will have been installed. The FAR construction contracts are summarized in Table 2.

Construction

14. Construction of the West Bank facilities commenced in December 1985 and is expected to continue through 1995. Over those 10 years, the value of the construction will be approximately US$552 million and £E423 million. The work has involved 12 contracts awarded to American contractors and 45 to Egyptian contractors. As of the start of 1993, 8 American and 12 Egyptian contracts had been completed. All contracting was executed with a process of prequalification for both American and Egyptian contractors carried out prior to actual tendering.

15. With one exception on the Giza Relief System, this arrangement has worked very well. In that case, which involved one of the earliest awarded contracts, it was necessary to split the work into two contracts. One, involving a major pumping station at Zenein, stayed with the American firm and the other, involving sewer construction in congested areas, was awarded to a local contractor. This modification resulted in both parts of the work being successfully completed.

16. The secondary and lateral sewers and service connections were awarded to local contractors because of the location of the work in small narrow streets and of the need to train such contractors in the installation of sewers to higher standards than had been adhered to in the past. With nearly 50% of these sewers in place, the decision to proceed in this way has proved very successful. At the start of this work the contractors struggled with both technical and logistical problems. Eventually, after the AMBRIC construction supervision team had provided considerable training in construction scheduling and techniques, the contractors improved their operations and performed very well. It is very likely that American contractors with large equipment-oriented operations would have been inappropriate anyway. This type of work calls for a labour-intensive approach. The result is that sewers of excellent quality are being installed under extremely difficult conditions and a large number of local contractors are now competent to undertake similar works. In addition, a large group of local engineers and inspectors has been trained in this work.

Major facilities

17. Of the 15 main contracts, 12 have been, or are nearly, completed, while three began construction in 1993. These contracts are listed in table 6 in reference 1.

18. The overall programme of construction of these contracts is provided in fig. 4 of reference 1. The contracts to construct the eight screw pumping stations were complete on or ahead of schedule. However, the two main contracts for the long lengths of multi-barrelled box culvert were delayed by the project-wide shortage of locally manufactured acid-resistant blue bricks.

19. Early difficulties were experienced in the commencement of sewer laying in the congested and densely populated Embaba area. American contractors initially withdrew, considering much of the work to be impossible owing to the presence of so many poorly constructed buildings and the high water table. The fears were proved to be unfounded, with the main collectors being constructed using the Kring method of shoring and trenching designed for work in soft soil. Combined with good dewatering techniques this method enabled the Contractor to make excellent progress on both contracts 20A and 27 in the Pyramids area. Pipe jacking of sewers up to 2750 mm was also successfully employed on lengthy road and canal crossings.

20. The Zenein treatment plant was the first to be commissioned, in October 1990, after a

complete rehabilitation. The start-up had been delayed for six months while repairs were made on the long sludge pressure mains from the plant to Abu Rawash, and while the permanent 66 kV power supply was provided.

21. As the existing sludge drying beds at Abu Rawash were being renovated as part of the contract to construct the new 0·4 million m³/day primary plant there, a temporary sludge-holding lagoon was built to retain the Zenein sludges.

22. Completion of the new Abu Rawash plant in October 1992 enabled the new system of pumping stations and culverts from Embaba to be commissioned. Flows were initially limited to 0·1 million m³/day until later in 1993 when the agricultural drains downstream of the plant will have been expanded and improved under contract 30.

23. The last major contracts on the West Bank scheme phase 1 will be the construction of a new steel pressure mains and pumping station to move all surplus Zenein and Abu Rawash sludges some 33 km to sludge storage lagoons in the Western Desert.

Construction challenges

24. The works may be divided into three very different types, all of which necessitated special methods and materials. The first type of work involved rehabilitation of the existing facilities throughout the West Bank, which were in extremely poor condition. Part of this rehabilitation required the construction of new pipelines, both rising mains and gravity lines. Many pumping stations were to be rehabilitated, but were then to be phased out on completion of the new works. The pumping stations and related pipelines were all located in congested areas and required careful attention in all aspects of the work. Ordering of replacement equipment was severely hampered by lack of data on the original procurement and installation. Drawings were virtually non-existent. In addition, much of the equipment had been kept in operation by cannibalization other similar but not identical equipment. Lastly, extraordinary effort and methods were required to keep wastewater moving while the pumping stations under construction were out of service.

25. The second type of work involved the construction of large collector sewer systems in highly congested urban areas and a system of eight low-lift (Archimedes screw pumps) pumping stations with large box culverts. Two of these pumping stations were built in congested urban settings, while the remainder, including the culverts, were constructed in and across open agricultural areas. A major effort

in dewatering was necessary for all of these works because normal groundwater levels were about 1·5–2 m below grade and all of the new works involved excavations in the 4–5 m depth range. Lastly a large (0·4 million m³/day) primary wastewater treatment plant, expandable to secondary and 1 million m³/day, was constructed at the extreme westerly limit of the project area. Although this facility was actually in the desert and about 25 km away from the Nile, groundwater levels were very high and they required a major dewatering effort. This involved deep wells which continuously pumped about 0·12 million m³/day of water over a two-year period.

26. The third type of work was the most challenging overall. It involved the construction of secondary sewers and service connections in highly congested districts (the FAR contracts). Many of the streets were only 4 m wide with excavations 3–4 m deep and groundwater at or near the surface. Extremely close attention needed to be given to dewatering operations, particularly regarding the pumping of fines. The buildings in these districts all had shallow foundations (1–2 m maximum). Prior to construction all buildings were closely inspected for any signs of structural distress. In many instances arrangements were made to evacuate the inhabitants and to provide for temporary housing. Another area of great challenge was in training the local contractors to plan their work in better ways and to meet high quality international standards of construction.

Egyptian antiquities

27. A significant part of the West Bank scheme was constructed in the area of Nazlet el Semanne village, adjacent to the major antiquity sites around the 4500 year old Pyramids and the Sphinx. Great care was taken in the planning of the sewerage systems in this area. Exploratory boreholes were sunk, not only to assess the nature of the ground along the sewer routes, but also to enable longer term monitoring of groundwater pollution on the Sphinx route outcrop.

28. Close monitoring of all excavations was carried out in conjunction with the Egyptian Antiquities Organization. This enabled rapid recording of significant finds and also enabled work to proceed with the minimum of delay and interruption. During the course of the contracts the most notable find was the remains of the previously undiscovered Valley Temple of Pharaoh Cheops. Archaeologists were also able to study and record excavations of the ancient village of the workers who constructed the Great Pyramids.

Reference
1. KELL A. D. K. *et al.* Project objectives, organization and implementation. *Proc. Instn Civ. Engrs, Civ. Engng, Greater Cairo Wastewater Project*, 1993, 8–17.

*Proc. Instn
Civ Engrs,
Civ. Engng,
Greater Cairo
Wastewater
Project,
1993, 60–63*

Paper 10238

Contract procedures

EurIng J. F. A. Barnard, BSc, FICE, FIWEM, ACIArb, *and*
R. J. Kachinsky, BSc, MSc, DEE, FASCE

■ Introduction

In 1982 a contract strategy report was prepared setting out the basis of implementation of the East Bank scheme. Topics which were covered included contract packaging, type of contract, conditions of contract, insurance and tender procedures. Similar considerations were given to the West Bank scheme, although these were not set out in a formal report. The detailed procedures adopted on each bank had some differences resulting from the requirements of the funding agencies and from variations between American and British practice.

2.   It was decided that the major construction contracts for tunnels, culverts, pumping stations and treatment works should be participated in by international contractors. On the East Bank the tunnel and pumping station contracts were restricted to UK—Egyptian joint ventures, the culvert contracts were restricted to Egyptian contractors with limited sterling being provided for UK construction management assistance, and the treatment works contracts were restricted to joint ventures from EC plant suppliers and Egyptian civil contractors. On the West Bank only American contractors were eligible to bid for the main contracts although they were permitted to subcontract part of the work to Egyptian contractors.

3.   The work on the West Bank also included a large group of smaller value contracts for the sewerage of substantial areas which previously had no public sewerage facilities. These were funded in Egyptian currency with assistance from grants provided by USAID for 79% of the tender value. These contracts are known as fixed amount reimbursable (FAR) contracts because the USAID funding element of the tender value was fixed. The contracts were offered for tender only to Egyptian contractors and, because one of USAID's objectives in providing the funding for the West Bank scheme was to help strengthen the private sector of the Egyptian construction industry, at least half of the contracts were restricted to private sector tenderers.

Conditions of contract

4.   In view of the international participation in the construction of the major contracts on both banks of the Nile, the FIDIC Conditions of Contract,[1] 3rd edition, were used for all these contracts. The contract documents contained numerous amendments to the basic FIDIC conditions to suit the requirements of the Egyptian authorities and of the funding agencies. The FAR contracts used a derivative of a local form of fixed-price contract.

5.   English is the governing language in all the FIDIC contracts but Arabic is the governing language in the FAR conditions of contract. The contract documents for FAR contracts present the conditions of contract in both languages; those for the FIDIC contracts were given only in English. All contracts are in accordance with and governed by Egyptian law. At the same time, under the tripartite funding protocol and by endorsement of the Egyptian Council of State, the contract provisions of such funded contracts are deemed to be part of Egyptian law.

6.   All of the FIDIC contracts include standard provisions for seeking a decision from the Engineer: AMBRIC, in the case of a dispute, followed by arbitration if necessary. For contracts with international contractors and UK funding any arbitration would be outside Egypt. Arbitration would be in Egypt for other contracts.

7.   Insurance provisions differ between the East and West Bank contracts. Following a study of likely benefits the Client, Cairo Wastewater Organization (CWO), decided to adopt an owner-arranged insurance providing cover for the Contractor, CWO and AMBRIC for most of the East Bank contracts. Contractors were provided with details of the policy and required to provide any additional insurance that they considered necessary. For the West Bank the normal FIDIC insurance provisions pertained.

8.   The clauses dealing with payment differed according to the funding agency requirements. The East Bank contracts, paid in pounds sterling and Egyptian pounds (£E), provide for some index-linked variation in prices, as do the FAR contracts paid in £E. However, the West Bank contracts which are paid wholly in US dollars, are essentially fixed-price. The period of construction has seen considerable exchange rate fluctuations between the three principal currencies used (Fig. 1). The corresponding movement of the local variation of price (VOP)

*J. F. A. Barnard,
Binnie & Partners;
Deputy Project
Director, AMBRIC*

*R. J. Kachinsky,
Camp Dresser &
McKee International
Inc.; West Bank
Manager,
AMBRIC*

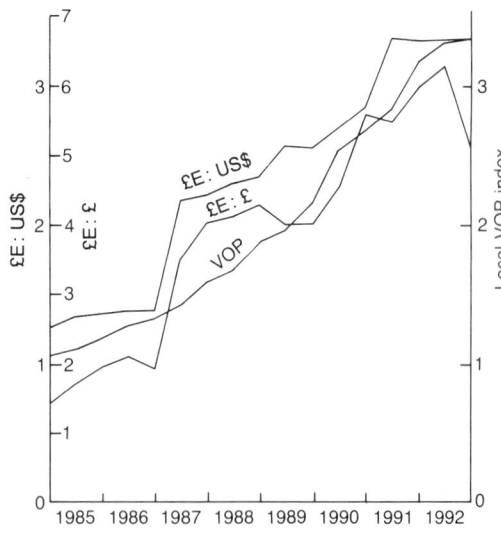

Fig. 1. Exchange rate and local variation of price index fluctuations

index used for the index-linked contracts is also shown in Fig. 1.

9. The payment procedures for West Bank FIDIC contracts allow amounts certified by the Engineer to be paid to contractors directly by USAID from grant funds if CWO does not process applications for payment within 40 days. However, the provisions for certification and payment in East Bank and FAR contracts require the Employer's approval of amounts certified before payment may be made in Egyptian currency from the Government of Egypt or USAID funds or in sterling from grants and loans available to CWO. Under Egyptian law, improper approval of public expenditure could be punished severely, and there arose a contractually undesirable but understandable reluctance by members of the Client's staff to approve payment certified by the Engineer until they had a full understanding of the details. The process of becoming assured that there was no contravention of internal regulations has led to some delayed and incomplete payments.

10. A noteworthy factor in the cash-flow problems on the East Bank arose from the effects of local taxes. All FIDIC contracts include a clause to exempt contractors from local taxes, duties and so on but, although exemption was obtained under the Government of Egypt–USAID agreement governing disbursement of USAID funds on the West Bank, no exemption was obtained for East Bank contractors. Hence reimbursement had to be arranged through certification, which increased the amount of finance needed by the Contractor and introduced extra work for the Engineer and Employer in dealing with Egyptian customs, tax laws and regulations. New fields for argument between the Contractor, the Engineer and the Employer were also opened and resulted in significant disallowances or delays in the payment of amounts certified.

11. The payment situation for Egyptian currency has been further worsened by chronic shortages of funds available to CWO, due to underallocation by the Ministry of Finance to budget applications from CWO. The combined effects of these cash-flow restrictions have been significant underpayments of Egyptian currency certified by the Engineer on most contracts. The amounts of the underpayments have varied between contracts and caused some contractors to claim that their ability to maintain their required rate of progress was affected.

12. Under the FIDIC conditions of contract non-payment could be used by the contractors as grounds for termination of the Contract. Otherwise a clause in part II of the conditions of contract provides for interest to be certified for underpayments. Therefore unless the Contractor could demonstrate that underpayment had indeed affected his ability to perform, AMBRIC awarded no extension of time. However, some of the FAR contracts include a provision, introduced during pre-award negotiations, for extensions of time equal to delays in payment of certified amounts. Substantial extensions to FAR contracts have resulted in many cases for this reason.

Specifications, bills of quantities and measurement

13. Specifications prepared for contracts forming part of the East and West Bank schemes differ in format according to British or American standard practice but, as the requirements for the project had been developed jointly in the early stages, the technical content is very similar. The technical requirements are incorporated into the specifications using the appropriate Egyptian standard where suitable, and elsewhere generally using British or American standards according to the source of funding.

14. Most of the West Bank contracts, based on common American practice, include only lump-sum-based bills of quantities with single items such as a complete screw-pumping station, supported by a schedule of prices to assist in variations and the evaluation of claims. Interim measurements for progress payments are on the basis of achievement of percentage completion of previously agreed elements of the works and are therefore comparatively simple, although approximate. The FAR contracts include items for sewer construction by linear metre according to depth plus items for house connections, manholes and similar work by complete unit. However, the East Bank contracts generally adopt the British practice of remeasuring detailed bills of quantities for such civil works items as concrete, reinforcement and shuttering, and similar detail for electrical and mechanical work. Consequently the preparation of progress payments requires more time and effort, but probably yields more precise interim evaluations.

Tender and award

15. Prequalification of tenderers for the contracts funded offshore was initiated by advertisement in the *Commerce Business Daily* for USAID-funded contracts and in the appropriate UK or other construction trade journals for other contracts. A questionnaire was sent to each responder and from the answers, together with further inquiries and clarification where needed, a short list of suitable contractors was drawn up by AMBRIC and submitted for approval by the Employer (CWO) and the consent of the appropriate offshore funding agency.

16. The tender period varied according to the value and nature of the work to be constructed and was extended for some contracts where necessary to ensure a sufficient number of competitive bids. In general, lively interest was shown in the tenders and bidding was close and competitive. On submission to CWO the tenders were opened and the prices announced at a public ceremony before they were passed to AMBRIC for assessment and ranking in a report to CWO and the funding agencies. A period of negotiation between CWO and the tenderers then followed and was sometimes protracted before a Contract was finally signed by CWO as the Employer. Except if funded in dollars by USAID, the contract would not come into force until advance payment had been made; in some instances where the Gov-

ernment of Egypt controlled the advance payment this added a lengthy period until commencement.

17. For the American contracts on the West Bank funded by grant aid the period between tender submission and the order to commence was generally at least 8 months and a maximum of 20 months. For the contracts on the East Bank funded by a combination of UK loan and Egyptian fu ıds the shortest period was 11 months and tl e longest 35 months.

18. The overall effect of the delays in award, combined with those during construction, was to exceed the allowance built into the original project programme and to require contingency provisions to be made to accommodate the staggered completion dates. On the West Bank the contingency provision was to arrange a contract for interim care and maintenance of those parts of the project which were complete but which could not be put into service because of delay to other parts—principally the treatment works. This continued until October 1992 when the main sewers, culvert, collector system and pumping stations were commissioned at the time the Abu Rawash treatment works became ready for service. On the East Bank the construction contracts already included such a provision, although it was of limited duration. To prevent this period being exceeded and to obtain the earliest possible benefit from the completed contracts, a temporary discharge arrangement was constructed from an intermediate pumping station. This allowed most of the East Bank scheme to be put into service in January 1992 despite lack of completion of the downstream elements of the complete system.

Construction value

19. Table 1 shows the number of contracts awarded by AMBRIC during the project up to the end of 1992. Of these, 34 are in the FAR category with values up to £2 million and totalling about £50 million. The values of other contracts [2] range between about £4 million and £100 million and total about £850 million. It is expected that 11 more contracts will be started during 1993. Of the total value certified to the beginning of 1993 the FAR contracts account for about £24 million and FIDIC contracts about £725 million. The ratio of equivalent value certified in Egyptian and foreign currency changed slowly from 1 : 2·9 in 1988 to 1 : 2·3 in 1992. Fig. 2 shows the total cumulative certified contract value to date on the project, and the programmed future expenditure on contracts still in progress or with funding available.

Contract finalization

20. During the course of contract administration and finalization 68 issues have been put to the Engineer for a decision under clause 67 of the conditions of contract. Up to the end of 1992 only two notices by a contractor of the intention to seek international arbitration under the FIDIC conditions had been lodged.

21. However, a number of arbitrations have been held within the Ministry of Housing to

Table 1. Contracts started and completed (up to 31 December 1992)

Year	Contracts started	Contracts completed
1984	5	—
1985	5	—
1986	3	2
1987	4	—
1988	2	3
1989	8	3
1990	11	8
1991	10	6
1992	16	14
Total	64	36

resolve disputes on payment or liquidated damages which involve contractors who are in the Egyptian public sector.

22. The contracts generally include a 12 month defect liability or maintenance period which, on the East Bank, only starts from initial operation or commissioning of the works rather than from the issue of the certificate of completion. In the case of electrical and mechanical plant there is also provision for 12 months of operation after major repair or replacement during the defect liability period. A residual responsibility for latent defects after expiry of defects liability periods continues under clause 63 of the FIDIC conditions.

23. In those East Bank contracts where commissioning has been delayed, the finalization of the contracts has thus also been delayed by potentially two years or more. Although undesirable from a contract administration point of view, the extension of the contracts to provide a defects liability period after commissioning has been of undoubted benefit to the Employer, particularly where mechanical and electrical plant has been involved.

24. Final Payment Certificates for completed contracts have been issued by the Engineer within the appropriate time from the issue of the Maintenance Certificates. However, in the case of contracts where final payment is controlled by the Employer, the full amount of the final payment has generally not been made within the due period. Instead, the Employer and Contractor have entered into negotiation to arrive at a figure for settlement. In this situation the Engineer has directed that the Final Payment Certificate will be amended only if an Engineer's decision on any element of it is formally requested by either party to the Contract pursuant to clause 67 of the FIDIC conditions of contract and, following arbitration if necessary, leads to an amended valuation, or if the parties notify the Engineer in writing that a settlement agreement has been reached in a different amount.

25. To date, no clause 67 disputes on any elements of Final Certificates have been notified to the Engineer for decision as a consequence of the Employer withholding payment; instead protracted negotiations have continued between the Employer and the Contractor on final payment for several contracts.

Staffing

26. Contract supervision on site is provided by an expatriate Engineer's Representative assisted by a team of expatriate and Egyptian technical and other staff. Powers under some clauses of the FIDIC conditions of contract are delegated by the Engineer to the Engineer's Representative. Others are retained at the level of overall management of services during construction on the East or West Bank as appropriate and yet others are retained within AMBRIC's central unit by the project director and his deputy. In the cases of clause 67 (settlement of disputes) and 52·5 (settlement of major claims) approval from the members of AMBRIC's Board of Control has to be obtained before powers can be exercised. The prior approval of CWO is required under AMBRIC's terms of reference before powers to order additional work can be exercised if the increase in contract sum would be substantial.

27. All aspects of construction, including quality control, progress, site health, safety and security, are the responsibility of the Contractor. The Engineer is responsible to the Employer for the monitoring of quality control by inspection, the witnessing of testing and sampling by the Contractor and, if need be, by independent testing. He has a monitoring and reporting role with respect to progress and contract cash flow. In the cases of site health, safety and security the Employer is the Authority within the Egyptian system and the Engineer is responsible for monitoring and reporting on the Contractor's compliance with regulations. Design and detailing is under the Engineer's responsibility and any alternative design proposal by the Contractor requires the approval of the Engineer before construction is started.

28. It is difficult to draw clear conclusions on staffing because of the variety of construction services provided and their changing nature with time. For instance, although the value of construction work per AMBRIC expatriate staff member has decreased from the equivalent of about £1·5 million per year in 1988 to about £1·3 million in 1992, this is because much of the work has been reduced in scale but involves more contracts. About 15 contracts were in progress in 1988 with an average certified value equivalent to about £6·2 million each during the year, whereas in 1992 there were 27 contracts in progress with an average certified value equivalent to about £2·6 million each. In addition an unidentifiable amount of expatriate time is taken up on transfer of knowledge and expertise to Egyptian staff and in discussions with the engineers employed by CWO on construction sites for liaison and observation. The substantial delays in completion of many contracts had a significant effect on the man-months needed for services during construction.

29. The overall average number of Egyptian and expatriate staff per contract has decreased from over 22 in 1988 to fewer than 14 in 1992. Throughout the project there has been a reduction of expatriate staff ratios during the same period from 1 per 3·6 total staff to 1 per 4·7. This is an indication of progressive transfer of responsibility to Egyptian staff together with the transfer of experience during the project.

Conclusion

30. The construction industry in Egypt was previously predominantly in the public sector, and many of the public-sector Employers have become used to a master–servant relationship with contractors rather than the equal relationship embodied in contracts such as the FIDIC conditions of contract. The custom was that the Contractor was required to carry out the Employer's requirements, and was reimbursed his costs from the public purse as necessary to maintain operating viability. If finance became temporarily unavailable the project would stop until such time as funds became available. The Contractor was responsible for quality and performance, and there was little or no independent check.

31. The process of learning and becoming accustomed to the obligations of the Employer, the Contractor and the Engineer under international commercial forms of contract is still continuing on this project. It has received added impetus from the policy being applied by the Government of Egypt, with encouragement from the World Bank and other agencies, to transfer much of the construction industry from the public to the private sector. The viability of the transfer will depend on the understanding gained, on this and other major projects, of commercial principles for construction contracts.

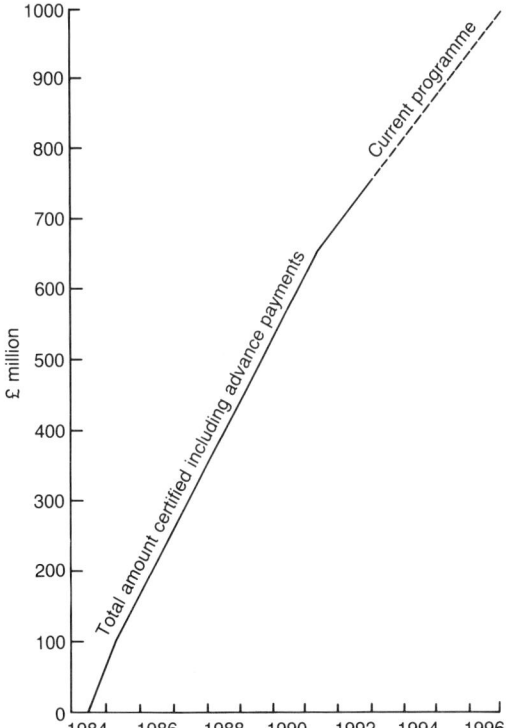

Fig. 2. Total construction contract expenditure

References
1. FEDERATION INTERNATIONALE DES INGENIEURS CONSEILS. *Conditions of contract for works of civil engineering construction*, 3rd edn. FIDIC, Lausanne, 1987.
2. KELL A. D. K. *et al.* Project objectives, organization and implementation. *Proc. Instn Civ. Engrs, Civ. Engng, Greater Cairo Wastewater Project*, 1993, 8–17.

Proc. Instn Civ Engrs, Civ. Engng, Greater Cairo Wastewater Project, 1993, 64–67

Paper 10243

Post-construction services

R. O'Kane, BSc, MSc, MICE, and J. D. Reece, BSCE, PE

■ **Introduction**

An important objective of the project was 'to properly operate and maintain all facilities' in order to protect the huge investments made by the Government of Egypt and the funding agencies and to prolong the useful life of the project works. This Paper describes training and other support given to the General Organisation for Sanitary Drainage (GOSD) both to improve the operation and maintenance of the existing sewerage system and to prepare for the commissioning and take-over of the main facilities. It also describes the comprehensive procedures necessary for the successful commissioning of the East and West Bank schemes.

2. The responsibility for the operation and maintenance of the sewerage system lies with GOSD. GOSD is a large public sector organization with approximately 11 000 employees. Its service area covers 921 km², including Helwan to the south of the project area. Within this catchment the new, existing and planned systems have a design sewered population of 13·5 million people (stage 1) with a treatment capacity of $3·3 \times 10^3$ m³/day by 1995.

3. Before commissioning of the project works, GOSD was responsible for the operation and maintenance (O&M) of an extensive sewerage system, including 120 km of large gravity collectors and force mains, over 4000 km of local sewer networks, and pumping facilities comprising 82 pneumatic ejector stations and 95 conventional pumping stations. Six primary treatment plants were also being operated, although these were heavily overloaded and were providing only rudimentary treatment. During the early 1980s GOSD was experiencing increasing difficulty in operating and maintaining the system. There were frequent reports of failures at pumping stations, sewage flooding incidents, and blockages and surcharging of sewers.

Early support programme

4. It was clear that the major rehabilitation programme for the existing GOSD facilities would be effective only if operational support was provided in parallel. An early priority was to improve the capacity of the existing sewers by the removal of accumulated grit and other deposits. During this phase, which was funded by USAID grants, several hundred GOSD operations staff were trained in the use of bucket cleaning and hydraulic jetting machines, in safety procedures and in sewer system maintenance. Additionally, as the rehabilitated pumping stations were commissioned, training programmes were implemented which covered basic operations and the preventive maintenance tasks required to keep the stations operational until the new project facilities could be commissioned.

5. By the mid 1980s construction work was well under way on the new project works and attention was shifted to the future management and O&M needs of the new facilities. A study of the strengths and weaknesses of GOSD was carried out in order to provide the framework and foundation for a broad programme of institutional strengthening. This study was completed by late 1989 and a four-volume report was prepared which comprised an action plan, a systems management plan, a systems operation plan and a comprehensive training plan. These plans analysed GOSD's existing managerial, organizational, financial and O&M capabilities, and recommended major strengthening and training initiatives throughout GOSD to enable it to function as a modern wastewater utility.

6. Over 600 GOSD personnel received training in a wide range of subjects including the O&M of pumping stations, treatment plants and sewer networks; computer processes and applications; laboratory process control; and management and training skills. A large amount of specialized equipment was also provided for GOSD which included computers, audio-visual equipment, laboratory and safety equipment, and tools.

Preparation for operation of the new schemes

7. In 1989 AMBRIC was commissioned to form a post-construction services (PCS) group to address the training, commissioning and O&M needs on the East and West Banks. On the West Bank, comprehensive PCS training programmes and O&M support initiatives were taken forward under USAID funding. On the East Bank the ODA funded O&M support programmes including an O&M contract for the major pumping stations. For the longer term, the organizational and management needs of GOSD are being addressed by the American contractor CH2MHill, under a major institutional support contract for GOSD, funded by USAID.

8. The groundwork for this initial phase for the new facilities was set out in reports prepared in 1990. These included detailed commissioning programmes highlighting the critical activities and responsible agencies. GOSD staffing needs and O&M costs were identified, organizational options studied and a comprehensive action plan was prepared to facilitate continuous monitoring of progress on all critical elements, including construction works, improvements in downstream drains, the provision of power supplies and provisions for sludge disposal.

9. A key element was the preparation of the system and facility manuals which form part of a comprehensive three-tiered O&M manual for the project works. On the first tier, the system

R. O'Kane, Acer Consultants Ltd; East Bank Post-Construction Services Engineer, AMBRIC

J. D. Reece, Black & Veatch International; West Bank Post-Construction Services Engineer, AMBRIC

manual sets out the operation of the overall collection and conveyance systems on each bank. On the second tier, a facility manual is provided for each pumping station within the system. The third tier comprises the manufacturers' detailed instruction manuals which were prepared by the contractors for each contract.

10. The timing of training programmes differed on each bank. On the East Bank training was included at Ameria, Kossous and Khalag pumping stations in an operation, maintenance and training contract, following commissioning in January 1992. On the West Bank a three-year training programme was carried out in advance of final West Bank commissioning in October 1992. These arrangements reflect the differences in philosophy and availability of funds of the funding agencies. The main focus of training provided for GOSD was on management and O&M skills specifically related to the new project facilities. The institutional problems associated with the GOSD organization, originally identified in the 1989 studies, were addressed with the appointment of CH2MHill to carry out a major programme of institutional strengthening for GOSD, under the Institutional Support Contract (ISC), funded by USAID. The goal of the ISC is to strengthen the managerial, organizational, technical and administrative system of GOSD and the scope of work embraces a comprehensive range of activities, including computerization and data management; financial viability and control systems; organizational development, policies and procedures; personnel management; maintenance management and inventory control; and safety and training. This programme will initially extend through to 1994 with further extensions if required.

O&M strategy and training programmes

11. On the East Bank, limited formal training services for GOSD staff were arranged before commissioning. These were carried out by plant suppliers as part of the contracts.

12. These arrangements were clearly insufficient and the ODA agreed to support a proposal to appoint an experienced O&M contractor to run the East Bank pumping stations and at the same time train GOSD staff until they were competent to assume full responsibility. Initially, O&M services were provided on an interim basis by the plant contractor, CEGELEC, and in August 1992—seven months after commissioning—full OM&T responsibility was taken over, following competitive tender, by a UK/Egyptian joint venture of Biwater Operations Ltd and Engineering Consultants Group.

13. The 24 month contract includes a comprehensive training programme specifically for the 236 GOSD staff for Ameria, Kossous and Khalag pumping stations. Training covers the operation, maintenance, administration and finance functions required to run a modern pumping station complex effectively. The focus is on job-specific hands-on training. The contract includes staged handover of the facilities to GOSD on successful completion of training.

14. For the West Bank scheme the OM&T strategy for GOSD included a training programme developed and implemented by AMBRIC, together with O&M training provided by some of the main contractors. The programme included training of GOSD's training department. To ensure that facilities were maintained correctly in the period between completion and operation, a maintenance contract for the screw pumping stations was also implemented.

15. The programme included workshop training in repair and maintenance skills, and general training in the areas of wastewater treatment, the operation of screw pumping stations, wastewater analysis and the management of facilities.

16. The workshop training which has already been provided for over 130 GOSD staff is focused on repair and maintenance in three areas: submersible pumps, diesel engines and electrical equipment. It is intended that the maintenance function of the West Bank screw pumping station will be centralized at Zenein and that mobile maintenance crews will service the seven stations.

17. Activities under the general training programme have included the training of operators and managers for the seven screw pumping stations and two treatment plants, the production of 26 training manuals for instructors and students, and the presentation of classroom and on-the-job training using these manuals. On-the-job training for the screw pumps was carried out at Pyramids station, which was specially modified for training purposes. Six 20 day courses have been provided for the 90 GOSD engineers and technicians needed for the seven stations. Additional operation training will be provided by GOSD instructors with technical advice from AMBRIC.

18. Training manuals have been produced in English and Arabic on a range of subjects including the operation of screw pumping stations and wastewater treatment plant (WWTP), WWTP solids disposal, laboratory operation, the maintenance of diesel engines, electrical controls and submersible pumps, and instructor training, management of facilities and safety. Over 1000 GOSD staff have received training in at least one of these subject areas.

East Bank commissioning

19. Delays in the start of Gabal El Asfar WWTP have meant that full commissioning of the complete East Bank conveyance system has had to be postponed until the second half of 1993. However, the construction contracts for the spine tunnel, culverts and pumping stations were complete in late 1991 and it was agreed that considerable relief could be provided to the overloaded sewers in central Cairo by proceeding with a limited commissioning. This plan involved making a limited number of connections to the spine tunnel and commissioning Ameria and Kossous pumping stations and the

culverts to Kossous. At Kossous a bypass channel was provided to allow up to 0.65×10^6 m^3/day of wastewater to be diverted to the nearby drain.

20. The need for a separate commissioning or flow diversion contract was also identified. The tunnel construction contracts had been completed but sewer connections to the tunnel could not be made because of delays in completion of the downstream works. It was therefore decided to place all the sewer tie-in work in a single contract, which was awarded to a local firm — Arab Contractors — in October 1991.

21. Detailed planning for first stage of commissioning started in 1991 with the preparation of an action list of critical activities and responsible agencies. Considerable effort was also expended on co-ordination with outside agencies including the Ministry of Public Works and Water Resources who were contracted by CWO to carry out extensive dredging and widening operations and bridge reconstruction work to improve the capacity of the downstream agricultural drains. The major element of this work involved a 60 km stretch of the Belbeis and Gabal El Asfar drains which was improved to accommodate wastewater flows of 1.6×10^6 m^3/day which will be discharged from Gabal El Asfar WWTP (stage 1) and Berka WWTP.

22. In mid 1991, a steering committee under the chairmanship of the Governor of Cairo was established to monitor and expedite progress towards commissioning, which was fixed for 21 January 1992. An 80 day count down programme was prepared, which highlighted all the critical activities. A tunnel connection sequence was agreed with GOSD, which balanced their priorities for the relief of flooding and surcharging in central Cairo with CWO's constraints in respect of minimum self-cleansing flows and downstream drain capacity. Detailed start-up and operation procedures were developed, which covered every element of the system to be commissioned. These included pre-start-up checks, penstock configurations, flow regimes and monitoring and communication instructions to be followed by the O&M contractor. Emergency procedures were also developed covering worst-case scenarios such as flooding of the downstream drains, which would have required diversion of flows back to the GOSD system, and collapse of the old collectors when surcharge pressures were removed.

23. The final countdown towards commissioning started in early January 1992 with the final inspection of the complete 12 km length of spine tunnel and the 6 km of twin and triple barrel culverts to Kossous. The first sewer tie-in was made at Ameria, which diverted collector 2 (0.275×10^6 m^3/day) to the collector pumping station. This allowed a controlled commissioning of the station and one barrel of the culvert system and permitted close monitoring of the downstream drain performance. After a satisfactory proving period of four days the Contractor was instructed to proceed with the tunnel connection sequence.

24. The first sewer tie-in to the spine tunnel was made 3.3 km from Ameria at shaft 4. This allowed 0.325×10^6 m^3/day to enter the tunnel providing sufficient flow to run one tunnel pumping station pump. This connection provided immediate relief to the overloaded sewers in the northern Shoubra and Shourabeya districts and allowed the major Souk El Samak pumping station to be decommissioned. Close monitoring of the flows in the tunnels, culverts and drains was maintained using mobile radios and the installed telemetry communication link between Ameria and Kossous.

25. Further sewer tie-ins were made to an agreed sequence and by the end of 1992 tunnel flow had built up to 0.8×10^6 m^3/day with an additional 0.2×10^6 m^3/day reaching the Ameria collector pumping station via the old collector system. The benefits to the sewerage networks in central Cairo have been dramatic, with sewage flooding eliminated and surcharging reduced in the catchments connected to the main tunnel. With the intended commissioning of the priority works at Gabal El Asfar WWTP in late 1993 and the completion of improvement works to the agricultural drains it is expected that the flow of 0.3×10^6 m^3/day currently diverted to the West Bank system via the Nile siphon will be admitted to the spine tunnel. Completion and commissioning of the Boulac branch tunnel and commencement of further branch tunnel construction are both scheduled during 1993. These developments will allow additional relief to overloaded East Bank networks and permit decommissioning of up to 50 pneumatic ejector stations and 30 conventional pumping stations.

26. The commissioning of the new facilities went remarkably well and the performance of the new conveyance system during its first year of operation has been in line with expectations. The flow regimes in the deep tunnel and culverts, and within the Ameria complex itself, have now been well tested. The high volumes of grit expected to enter the tunnel initially were successfully intercepted in temporary grit traps located at the vortex drop shafts into the tunnel. Although considerable volumes will continue to enter the system, these are moving through the new facilities as planned.

West Bank commissioning

27. Early commissioning of the renovated Zenein WWTP, with a capacity of 0.3×10^6 m^3/day, began on 7 October 1990 with the start-up of the first module. The second and third modules came into operation later in 1990. Initially the plant underwent trials to establish the most satisfactory mode of treatment. The plant, which is the first in Egypt to treat to secondary treatment standards, has consistently produced a satisfactory effluent from an influent of 262/332 BOD/SS. The plant was designed to produce a 30/30 effluent. Table 1 summarizes plant performance since 1990. The operation of Zenein WWTP has seen a gradual transfer of O&M skills from the expatriate contract staff to GOSD staff over two years. A final

Table 1. Performance of Zenein wastewater treatment plant (figures in mg/l)

	Influent		Effluent	
	BOD	TSS	BOD	TSS
October 1990	267	362	29	26
December 1990	310	483	29	43
January 1991	311	443	28	41
April 1991	361	347	40	23
July 1991	260	340	22	16
October 1991	242	348	20	11
January 1992	293	387	22	17
April 1992	261	323	17	11
July 1992	213	312	19	8
October 1992	233	292	23	10
January 1993	288	328	15	16

one year period of technical and advisory support will be completed in late 1993.

28. Commissioning of the north-west conveyance systems between Boulac and Abu Rawash began on 25 October 1992 following pre-commissioning tests at the completed 0.4×10^6 m^3/day primary treatment plant at Abu Rawash. Following a similar strategy to that employed on the East Bank scheme an initial flow of 85 000 m^3/day was introduced into the new sewers flowing to Boulac pumping station by diverting wastewater from existing systems. In early November the Embaba pumping station was added to the network.

29. Start-up of the new West Bank scheme has gone well; there have been only a few problems associated with minor component failures. All screw pumps and associated equipment have performed up to expectations and the conveyance system is now being operated by GOSD staff. At present the flows into the new system are limited to 100 000 m^3/day, but this constraint will be removed when the drain capacity is improved and constrictions downstream of Abu Rawash are removed. The Pyramids area will be connected to the system by the end of 1993.

Conclusions

30. Comprehensive training and other assistance has been given and is continuing to ensure that the GOE's massive investment to renovate and extend Cairo's sewerage system can be safeguarded by proper management, operation and maintenance of the system. The first stages of commissioning of the East and West Bank schemes have been completed successfully, and operation and maintenance of the newly commissioned facilities partly by GOSD and partly by OM&T contractors is proceeding satisfactorily.

Proc. Instn
Civ Engrs,
Civ. Engng,
Greater Cairo
Wastewater
Project,
1993, 68–72

Paper 10239

Conclusions and future requirements

J. P. Somerville, BSc, MSc, FICE, FIWEM, *and*
A. D. K. Kell, BSc(Eng), MSc, DIC, MICE, MIWEM

J. P. Somerville,
Projects Group
Manager, Acer
Consultants
Limited; Board of
Control and
former Project
Director,
AMBRIC

A. D. K. Kell,
Partner,
Binnie & Partners;
Board of Control
and former
Project Director,
AMBRIC

■ Introduction

By 1993 over £1 billion had been expended to rehabilitate Cairo's existing wastewater works, to provide main sewers and sewage treatment facilities and to bring first time sewerage to extensive areas on the West Bank of the River Nile. Substantial parts of the new system had been commissioned and construction contracts for the remainder were underway. Contracts had also been let to provide for the operation or interim care of facilities as commissioning of the system progressed and a comprehensive programme of training of the General Organisation for Sanitary Drainage (GOSD) operating staff was underway.

2. Many criteria can be used to assess the project's achievements including those of engineering, cost, programme, public health and socio-economics. However, in terms of the fundamental objectives of relieving the overloading of the existing system and the provision of new facilities able to provide for future growth in a way that facilitates incremental augmentation and environmental improvement, the project can be considered to have achieved its objectives.

Project evaluation

Overview of achievements

3. It was not part of the original planning for the project to cover either the rehabilitation of existing secondary sewerage areas or to provide first-time sewerage at areas not yet served by the system. This work was felt to be within the capability of GOSD and was included in GOSD's programme. However, it was decided to include within the programme for rehabilitation of the major facilities a demonstration scheme for the rehabilitation of the secondary sewerage network in an area in South Cairo, subject to continued flooding, known as the Abu Saud. Documents were prepared in 1981 and construction was carried out during the period 1983–86.

4. On the West Bank, substantial and increasing areas were not connected to the sewerage system. In 1985/86 a demonstration study was carried out to investigate the costs and effectiveness of a number of different methods of improving sanitary conditions in those areas, pending the provision of conventional water-borne sewerage. These included the trial of vacuum tankers for emptying vaults, the provision of communal vaults, communal latrines and vault connections to a sewerage system. While a number of these measures were technically feasible and were cost effective, it proved difficult to arrange an implementation programme as there was no Egyptian authority with the remit or the capa-

bility to take charge of it. Instead the Government of Egypt, with a promise of additional funding from the USA, decided to accelerate provision of water-borne sewerage. The FAR programme was put in hand in 1988 and, when complete in 1994, it will have provided sewerage facilities to an area of 1500 ha with a present population of 1·8 million. Fig. 1 shows the increase in sewered population and treatment capacity in Greater Cairo since 1975.

Effectiveness of implementation procedures

5. There are two main difficulties in evaluating the performance of the project in financial terms: (*a*) the effects of inflation over a period of more than a decade, and (*b*) the effects of the large variation in exchange rates which, in the case of the rate between £E and £, has varied by a factor of over five.

6. Funding from foreign sources has been for amounts which were fixed at the time of commitment. Any delay in the expected drawdown of the funds has reduced the value of the contribution due to the effects of inflation. The Egyptian government's contribution is set within the context of Egypt's five-year plan, but is allocated on a yearly basis according to updated estimates. Table 1 shows the financial performance of the East Bank scheme.

7. The organizational arrangements for the project are described in reference 1. At its peak, the project directly involved 400 staff from the Client, the Cairo Wastewater Organisation (CWO), 100 expatriates and over 400 Egyptians from the Engineer, AMBRIC and Egyptian Associates and over 6000 contractors' staff on the various construction sites. Many thousands more have been employed indirectly in the pro-

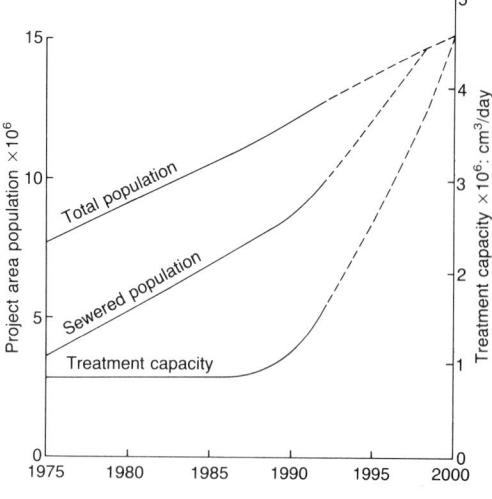

Fig. 1. Increase in sewered population and treatment capacity

Table 1. Financial performance of the East Bank project (figures in millions)

Contract number	Contract strategy		Contract award			Out-turn cost*		
	£	£E	£	£E	Year	£	£E	Year of completion
Tunnels								
Spine tunnels 3, 4 and 12	63·4	148·9	119·9	202·9	1984/1986	114·6	193·3	1988/1989
Branch tunnel 12/5A	NA	NA	49·0	129·8	1990	47·9	137·3	1993/1994
Tunnel connections 17	NA	NA	—	19·6	1991	—	19·6(E)	1993
Ameria Pumping Station Complex 1 and 2	25·4	28·9	26·0	25·0	1984	32·6	37·5	1991
Conveyance system 6, 7, 8 and 9	48·7	137·6	30·4	167·8	1984/1987	31·3	192·9	1990/1992
Treatment plant 16·1 and 16·2	NA	NA	64·3†	421·6	1991	64·3(E)§	421·6(E)	1993/1995
Totals			289·6	966·7		290·7	1002·2	

E = estimate, NA = not applicable. Exchange rate 1982 £E1·52 = £1·00, March 1993 £E4·71 = £1·00. Inflation 1982–92: UK 23%, Egypt 190%.
* £E figures include payments under clause 73 (variation in price and interest) and for interest. These were not allowed for in contingencies.
† Includes no contingencies; Contract awarded in US$96·5M.
§ No provision for claims.

vision of materials, plant and services to the project. A number of factors have contributed to the smooth running of the project and a co-operative approach to problem solving. Firstly, the location of the Client and the Engineer in the same building has promoted close liaison and co-operation between the two; secondly, the fully integrated organization of AMBRIC enabled the consultant to draw on the resources and expertise of Egyptians, Americans and British in an effective way; and finally, the close interest of the major funding agencies, the ODA and USAID, both on a day-to-day basis and through their participation at the regular liaison meetings in Cairo, helped in the steering of the project. As ODA, in contrast to USAID, did not have a full-time presence in Cairo, the British Ambassador and his staff played a major part in ensuring that any problems affecting the United Kingdom's performance were brought to the appropriate level of the Egyptian government.

8. The powers given to the Engineer in terms of certification and the ordering and valuation of changes were quite different from normal Egyptian procedures and caused many problems, especially in the initial stages. The obligation of the Engineer to give impartial decisions was also a matter viewed with suspicion by the Client although, in fact, of the 68 Engineer's decisions given to date only two have been referred to arbitration by the Contractor and none by the Client. A series of training courses for the Client's staff was arranged to explain the philosophy and workings of the FIDIC conditions of contract. These were helpful in resolving many of the misunderstandings that had arisen.

Performance of facilities

9. Experience of the functioning of the early sewerage schemes for Cairo showed problems with the deposition of sediment in the sewers and severe corrosion of concrete. The project has taken advantage of recent advances in sediment transport theory for the design of a system in which sewage, with a sediment load of up to 200 mg/l could be carried without build-up of sediment deposits. Experience in the first 12 months following commissioning indicates that the hydraulic performance has conformed with expectations. Substantial relief has been provided to the existing system and no build-up of sediment within the new sewers has been detected. The new system has been in operation for insufficient time for any conclusions to be drawn on the effectiveness of the corrosion protection provisions, but nothing untoward has been observed.

10. On 12 October 1992 Cairo was struck by an earthquake with a reported magnitude between 5·5 and 5·9 on the Richter scale. There was widespread damage in the older part of the city. A careful inspection, where possible, of the project works both in operation and under construction revealed a small amount of superficial damage. To date it has not been possible to inspect the interior of the spine tunnel which is in operation.

Sustainability of facilities

11. There is growing recognition that for a project to be judged a success, not only must it fulfil its objectives at the time it is commissioned and handed over to the owner, but also these objectives should be sustained in the long term.

12.   A number of factors affect a project's sustainability; the process starts during planning and design and continues through construction and the preparations for operation.

13.   At the planning and outline design stage of the project, particular attention was paid to the following factors which were judged to be important for the Cairo situation.

(a) Flows to be by gravity with the minimum number of pump stations and mechanical plant.
(b) There are a limited number of wastewater treatment works of large capacity.
(c) Tunnels and culverts have reserve structural strength in all conditions; corrosion protection is matched to the overall design life and minimum maintenance requirements.

14.   At the detailed design stage, great care was taken in the detailing and the specifications to ensure a robust design with the strength and corrosion resistance to ensure a long life and low maintenance requirements. Risk analysis was carried out to ensure, as far as possible, that standby facilities were available and that flows could be diverted or overflowed safely. Key installations were provided with a dual supply of electricity from the electricity authority, plus standby generation capacity sized to keep essential elements operational. A comprehensive inventory of spare parts was provided as part of the plant supply contracts, with particular attention to the adequacy of long delivery items.

15.   The vital importance of the training of operational staff was recognized early on. It included English language skills, classroom lectures, workshop training, overseas courses and placements, and on-the-job training at specific facilities. This training did much to improve the capability of GOSD to operate and maintain the new facilities but did not cover institutional matters in any depth. A programme of institutional development has now been put in hand with technical assistance from a separate consultant funded by USAID and working directly with GOSD. One matter of concern in ensuring the sustainability of the project is the need to retain operational staff once they have been trained. As GOSD is subject to the Egyptian government's normal civil service rules, it can only pay the prescribed government service rates. Staff trained to high standards of competence under the project enhance their chances of employment elsewhere in Egypt or in the Gulf States, and the training programmes have had to take account of the considerable wastage that occurs.

Evaluations by funding agencies
16.   In 1992 both ODA and USAID have carried out evaluations of the project by teams of professionals commissioned specifically for the purpose. A major objective of these evaluations was to satisfy their internal requirements of ensuring that the public funds allocated for the project have been used in a productive and cost effective manner. In addi-

tion, both evaluations have sought to assess the impact of the project on those benefiting from it. Social, institutional, environmental, economic and technical aspects were covered. The ODA also intends to use the feedback from the evaluation to strengthen its future project design and implementation strategy.

17.   Any attempt to quantify the benefits of the project is hampered by the paucity of reliable baseline data, particularly in relation to public health. Also any improvements in public health, if these could be measured and correctly ascribed, will take some time to manifest themselves following the commissioning of the facilities. The funding agencies are now considering how best to establish a baseline condition for public health in the city and how improvements can be measured in a few years' time.

Future requirements
18.   The Government of Egypt took its first steps to implement the Greater Cairo Wastewater Project in 1975. For various reasons, it was not until 1984 that construction work started. By 1993 major works had been commissioned and the government had gone a long way towards achieving its goals. However, in the intervening period the population of Cairo had increased by four million (63% more than the 1975 population), and it is still increasing at around 300 000 a year. In view of the limited availability of funds which could be allocated each year, the government first concentrated on eliminating flooding in the streets of Cairo, then provided sufficient additional conveyance capacity to meet future demands, together with enough capacity in the drains for disposal, and finally turned to the provision of treatment facilities.

19.   The provision of infrastructure facilities in a major and growing city can never be said to be complete. The Cairo authorities have overcome a major backlog and have provided some capacity for the future but there are still many tasks to be carried out before the system meets modern standards and the aspirations of the authorities in all respects. MOHU through CWO and GOSD are now turning their attention to the following matters.

20.   With the works in hand nearly all elements of the originally planned phase 1 programme have been completed or are underway. The major outstanding works for which funds are still to be committed are the remaining East Bank branch tunnels and the extension of the spine tunnel to Maadi in the south of the city. The European Investment Bank (EIB) is providing funds for the branch tunnels. Further consideration needs to be given to the timing for the Maadi tunnel as temporary pumping arrangements made by GOSD may have sufficient capacity to allow deferment of the tunnel construction for an additional period.

21.   The commissioning of the spine tunnel and the new pumping stations at Ameria has allowed the overloaded old collectors to be drawn down to levels which permit inspection for the first time in many years. The collectors

are known to have considerable deposition of silt and in some areas to have severe corrosion in the crown.

22. Advantage should be taken of the opportunity to clean out the sediment deposits and to carry out detailed condition surveys to determine the repairs that are necessary, so that an assessment can be made as to whether repairs would be cost effective.

23. In 1991 AMBRIC carried out a system load review. Its purpose was to compare predicted and attained flows and strengths of wastewater and to estimate, with the use of existing and readily available information, the loading which would be accepted by the new facilities as they were brought into commission and to determine whether there would be an early need for any of the facilities to be augmented (as provided for in the master plan). The review indicated that, subject to more detailed checks, a number of augmentation works will need to be put into hand in the near future. Table 2 shows which works are needed and the estimated date they will be required.

24. While the system load review was useful in providing an indication of requirements in the near future, there has been a growing awareness that the last full master plan was prepared almost two decades ago. In the intervening period the population of the city has increased by 87% and profound physical and social changes have taken place. Knowledge and awareness of the environment have increased considerably since the master plan was prepared, and reviews of environmental laws and standards are in hand in the general charge of the recently formed Egyptian Environmental Affairs Agency. It is clearly an appropriate time to prepare a new master plan taking account of the latest demographic, land-use and water supply data, physical plans for the future, and new strategies and regulations for the protection of the environment. The plan would be able to set the direction and scale of improvements over the next 25 years of development of the city.

25. One of the problems being addressed by the Egyptian government is that of inefficiency and stagnation within the public sector. A high-level Presidential committee is assessing the potential benefits of privatization and commercialization within certain sectors. While priority is likely to be given to privatization of industry and tourism, consideration is also being given to problems in infrastructure. Major difficulties are the attraction of good calibre staff and the retention of trained staff at the levels of salary which organizations such as GOSD are permitted to pay. Options under consideration include the re-establishment of GOSD so that it can increase levels of pay and hence attract and retain better quality staff, and the subcontracting by GOSD to private sector companies of operation and maintenance services. A comprehensive study of the various options needs to be undertaken to ensure the sustainability of the new facilities. This should include an estimate of the costs of the options, together with an assessment of the ability of the community to pay for them, methods of

Table 2. Expansion works

	Facility	Estimated year for implementation
East Bank	Ameria TPS	2005
	Ameria to Kossous PS: culverts	2000
	Kossous PS	1995
	Khalag PS	1995
	Kossous to Gabal el Asfar: culverts	1995
	Gabal el Asfar WWTP	1995
	Shoubra el Kheima WWTP	2011
	Berka WWTP	2001
West Bank	Pyramids PS	1995
	Junction PS	1995
	Abu Rawash PS	1995
	Junction–Abu Rawash: culverts	1998
	Abu Rawash WWTP	1993/1995
	Zenein WWTP	2003

charging for the services and the development of a tariff policy.

26. In 1991–1992 Taylor Binnie & Partners carried out an industrial pollution control study for CWO, who were assisted in funding the study by the EIB. It was agreed that a high-level steering committee should be set up to oversee the implementation of a number of measures including

(a) the review of laws and regulations relating to industrial waste discharges,
(b) the provision of assistance to industry in pretreating its wastes, and
(c) the setting up of an Effluent Management General Department with GOSD which would have the responsibility and powers to take and analyse trade effluent samples, impose limits on the discharge of substances, levy charges on discharges and impose penalties for offences.

The proposals form the basis for effective control of industrial discharges and should be put in hand without delay.

27. The conservation and efficient use of water resources is vital to Egypt's wellbeing and a key long-term objective of the Egyptian government is the reuse of sewage effluent. When the envisaged development of Cairo's five wastewater treatment plants is complete, they will be discharging over 5 million m^3/day (1·8 billion m^3/year). If the effluents were reused for agricultural irrigation it could be sufficient, depending on the crops to be grown, to irrigate over 100 000 ha of land.

28. Various studies have been carried out at prefeasibility level to assess the potential for reuse, the options available and their economics. The main alternatives are the irrigation of existing cultivated lands, probably using the irrigation drainage system to convey the treated effluent to the area to be irrigated, and the reclamation of desert lands which would need the creation of a new irrigation and drain-

age system. While desert reclamation appears attractive on many counts, the areas that could be reclaimed are generally at a high elevation compared to the treatment plants and of limited agricultural potential. Preliminary estimates indicate that the economic viability of such schemes is marginal.

29.   With the discharge of treated effluent suitable for agricultural reuse now a reality, and with large increases in the quantities available in the next few years, in-depth studies should now be put in hand to determine how to make the most effective use of this valuable resource.

30.   It has been estimated that over 3000 t of organic sludge will be generated at Cairo's plants each day. Sludge has potential for reuse as a soil conditioner, and has some fertilizer value. A sludge management study was carried out by AMBRIC in 1991. This indicated that there is a limited potential for reuse of the sludge and that the remainder will need to be disposed of to lagoons and landfill. The strategy now needs to be developed to confirm the potential for reuse, to develop an implementation plan and to identify suitable disposal sites and make appropriate arrangements for disposal as the quantities of sludge generation increase.

Reference

1. KELL, A. D. K. *et al*. Project objectives, organization and implementation. *Proc. Instn Civ. Engrs Civ. Engng, Greater Cairo Wastewater Project*, 1993, 8–17.